Lecture Notes in Computer Science 14121

Founding Editors

Gerhard Goos
Juris Hartmanis

Editorial Board Members

The series Lecture Notes in Computer Science (LNCS), including its subseries Lecture Notes in Artificial Intelligence (LNAI) and Lecture Notes in Bioinformatics (LNBI), has established itself as a medium for the publication of new developments in computer science and information technology research, teaching, and education.

LNCS enjoys close cooperation with the computer science R & D community, the series counts many renowned academics among its volume editors and paper authors, and collaborates with prestigious societies. Its mission is to serve this international community by providing an invaluable service, mainly focused on the publication of conference and workshop proceedings and postproceedings. LNCS commenced publication in 1973.

Mario Vento · Pasquale Foggia ·
Donatello Conte · Vincenzo Carletti
Editors

Graph-Based Representations in Pattern Recognition

13th IAPR-TC-15 International Workshop, GbRPR 2023
Vietri sul Mare, Italy, September 6–8, 2023
Proceedings

Editors
Mario Vento 🆔
University of Salerno
Fisciano, Italy

Pasquale Foggia 🆔
University of Salerno
Fisciano, Italy

Donatello Conte 🆔
University of Tours
Tours, France

Vincenzo Carletti 🆔
University of Salerno
Fisciano, Italy

ISSN 0302-9743   ISSN 1611-3349 (electronic)
Lecture Notes in Computer Science
ISBN 978-3-031-42794-7   ISBN 978-3-031-42795-4 (eBook)
https://doi.org/10.1007/978-3-031-42795-4

Preface

This volume contains the papers presented at GbR 2023: 13th IAPR-TC15 Workshop on Graph-based Representations in Pattern Recognition, held on September 6–8, 2023 in Vietri sul Mare, Italy.

GbR 2023 was the thirteenth edition of a series of workshops organized every two years by the Technical Committee #15 of the International Association for Pattern Recognition (IAPR). This was the first edition held after a hiatus due to the CoViD pandemic and the consequent travel restrictions; since the direct, informal interaction between the participants has always been one of the key aspects of this workshop series, the TC15 decided to skip the organization of the 2021 edition, rather than transform it into an online event.

This workshop series traditionally provides a forum for presenting and discussing research results and applications at the intersection of pattern recognition and image analysis on one side and graph theory on the other side. In addition, given the avenue of new structural/graphical models and structural criteria for solving computer vision problems, the GbR 2023 organization encouraged researchers in this more general context to actively participate in the workshop. Furthermore, the application of graphs to pattern recognition problems in other fields such as computational topology, graphic recognition systems and bioinformatics is also highly welcome at the workshop.

The volume contains 16 papers, of the original 18 submissions. Each accepted paper was single-blind reviewed by three program committee members. The program of GbR 2023 also included two talks by IAPR invited speakers: Walter Kropatsch (Professor Emeritus at the Vienna University of Technology, Austria) and Francesc Serratosa (Universitat Rovira i Virgili, Spain).

We want to thank the International Association for Pattern Recognition for making GbR 2023 an IAPR-sponsored event. We also thank the University of Salerno and the Department of Information and Electrical Engineering and Applied Mathematics (DIEM) for sponsoring the workshop.

September 2023

Mario Vento
Pasquale Foggia
Donatello Conte
Vincenzo Carletti

Organization

General Chairs

Mario Vento   University of Salerno, Italy
Pasquale Foggia   University of Salerno, Italy
Donatello Conte   University of Tours, France

Program Committee Chair

Vincenzo Carletti   University of Salerno, Italy

Web and Publicity Chair

Luca Greco   University of Salerno, Italy

Program Committee

Sebastien Bougleux   Université de Caen Normandie, France
Luc Brun   ENSICAEN, France
Ananda S. Chowdhury   Jadavpur University, India
Pasquale Foggia   University of Salerno, Italy
Benoit Gäuzère   LITIS – INSA de Rouen, France
Rosalba Giugno   University of Verona, Italy
Edwin Hancock   University of York, UK
Xiaoyi Jiang   University of Münster, Germany
Walter G. Kropatsch   Vienna University of Technology, Austria
Cheng-Lin Liu   Chinese Academy of Sciences, China
Josep Llados   Autonomous University of Barcelona, Spain
Bin Luo   University of York, UK
Jean-Yves Ramel   Université de Tours, France
Romain Raveaux   Université de Tours, France
Kaspar Riesen   University of Bern, Switzerland
Francesc Serratosa   Universitat Rovira i Virgili, Spain
Salvatore Tabbone   Université de Lorraine, France
Ernest Valveny   Autonomous University of Barcelona, Spain

Mario Vento   University of Salerno, Italy
Richard Wilson   University of York, UK

Local Committee

Antonio Greco   University of Salerno, Italy
Pierluigi Ritrovato   University of Salerno, Italy
Alessia Saggese   University of Salerno, Italy

Abstracts of Invited Talks

From LBP on Graphs to Slopes in Images

Walter G. Kropatsch ⓘ

Pattern Recognition and Image Processing Group, TU Wien, Austria
krw@prip.tuwien.ac.at

Abstract. Local Binary Patterns (LBP) are efficient texture descriptors by a binary code comparing the differences of grey values between a center pixel and its neighbors. This works well if the number of neighbors is constant (i.e. 4 or 8) like in digital images.

We introduce an equivalent code on graphs that allows the vertices to have different degrees. LBPs determine critical points (minima, maxima, saddles) without explicit differentiation even on plane graphs and the known drawbacks of differentiation in the presence of noise. Maximal monotonic profiles (1D curve) connect a local minimum with a local maximum while the 1D LBP between two successive extrema is always the same. A bounded range of values causes a long profile to have lower (average) inclination corresponding to a low contrast in images.

We propose to build a graph pyramid by successively contracting edges with lowest contrast and preserving critical points as much as possible. In contrast to most previous pyramid constructions these selections preserve not only the grey value range and the critical points but more importantly the high frequencies corresponding to the remaining high contrasts. Images can be reconstructed from a high pyramid level by simple inheritance from parents to childs. It is surprising how difficult it is to visually see the difference between the original image and a reconstruction using only a small percentage of colors/grey values.

We further explore the scientific question of what characterizes the neighborhoods of the critical points at a high pyramid level. We define slopes as regions of the image domain where every pair of points can be connected by a monotonic path. Such a slope can contain a single local maximum and a single local minimum but no saddle point strictly inside its domain. Saddles appear exclusively along the boundaries of slopes. The diversity of slope regions allows to partition images beyond internal and external contrast.

Several further interesting properties of slopes will be addressed during the talk.

Face to Face: Graphs and Biotechnology

Francesc Serratosa (iD)

Research Group ASCLEPIUS: Smart Technology for Smart Healthcare, Departament
d'Enginyeria Informàtica i Matemàtiques, Universitat Rovira I Virgili, 43k007
Tarragona, Catalonia, Spain
francesc.serratosa@urv.cat

Abstract. A chemical compound is basically a structural element. Nevertheless, until some years ago, non-structural representations, such as vectors, were used in machine learning applications, which predict compound toxicity or drug potency.

Currently, several machine learning methods have appeared in biotechnology that are feed by structural representations of chemical compounds, such as attributed graphs. This tendency change has emerged due to the appearance of more powerful machine learning methods boosted by the increase of computational power of servers and personal computers.

In this talk, several methods are going to be commented, ranging from the first graph matching methods to some of the latest tendencies related to graph machine learning. These methods have been applied in biotechnology to fields such as chemical compound classification, drug potency prediction and nanocompound toxicology prediction. More specifically, the aim is to remember and present old methods such as graph edit distance, K-nearest neighbours or graph embedding. But also, the aim is to comment the current methods such as graph convolutional networks, graph autoencoders, graph regression or saliency maps.

Contents

Graph Kernels and Graph Algorithms

Quadratic Kernel Learning for Interpolation Kernel Machine Based Graph
Classification 3
    Jiaqi Zhang, Cheng-Lin Liu, and Xiaoyi Jiang

Minimum Spanning Set Selection in Graph Kernels 15
    Domenico Tortorella and Alessio Micheli

Graph-Based vs. Vector-Based Classification: A Fair Comparison 25
    Anthony Gillioz and Kaspar Riesen

A Practical Algorithm for Max-Norm Optimal Binary Labeling of Graphs 35
    Filip Malmberg and Alexandre X. Falcão

An Efficient Entropy-Based Graph Kernel 46
    Aymen Ourdjini, Abd Errahmane Kiouche, and Hamida Seba

Graph Neural Networks

GNN-DES: A New End-to-End Dynamic Ensemble Selection Method
Based on Multi-label Graph Neural Network 59
    Mariana de Araujo Souza, Robert Sabourin,
    George Darmiton da Cunha Cavalcanti,
    and Rafael Menelau Oliveira e Cruz

C2N-ABDP: Cluster-to-Node Attention-Based Differentiable Pooling 70
    Rongji Ye, Lixin Cui, Luca Rossi, Yue Wang, Zhuo Xu, Lu Bai,
    and Edwin R. Hancock

Splitting Structural and Semantic Knowledge in Graph Autoencoders
for Graph Regression 81
    Sarah Fadlallah, Natália Segura Alabart, Carme Julià,
    and Francesc Serratosa

Graph Normalizing Flows to Pre-image Free Machine Learning
for Regression 92
    Clément Glédel, Benoît Gaüzère, and Paul Honeine

Matching-Graphs for Building Classification Ensembles 102
    Mathias Fuchs and Kaspar Riesen

Maximal Independent Sets for Pooling in Graph Neural Networks 113
    Stevan Stanovic, Benoit Gaüzère, and Luc Brun

Graph-Based Representations and Applications

Detecting Abnormal Communication Patterns in IoT Networks Using
Graph Neural Networks 127
    Vincenzo Carletti, Pasquale Foggia, and Mario Vento

Cell Segmentation of *in situ* Transcriptomics Data Using Signed Graph
Partitioning 139
    Axel Andersson, Andrea Behanova, Carolina Wählby, and Filip Malmberg

Graph-Based Representation for Multi-image Super-Resolution 149
    Tomasz Tarasiewicz and Michal Kawulok

Reducing the Computational Complexity of the Eccentricity Transform
of a Tree 160
    Majid Banaeyan and Walter G. Kropatsch

Graph-Based Deep Learning on the Swiss River Network 172
    Benjamin Fankhauser, Vidushi Bigler, and Kaspar Riesen

Author Index 183

Graph Kernels and Graph Algorithms

Quadratic Kernel Learning
for Interpolation Kernel Machine Based
Graph Classification

Jiaqi Zhang[1], Cheng-Lin Liu[2,3], and Xiaoyi Jiang[1(✉)]

[1] Faculty of Mathematics and Computer Science, University of Münster,
Einsteinstrasse 62, 48149 Münster, Germany
`xjiang@uni-muenster.de`
[2] National Laboratory of Pattern Recognition, Institute of Automation of Chinese
Academy of Sciences, Beijing 100190, People's Republic of China
[3] School of Artificial Intelligence, University of Chinese Academy of Sciences,
Beijing 10049, People's Republic of China

Abstract. Interpolating classifiers interpolate all the training data and
thus have zero training error. Recent research shows their fundamental importance for high-performance ensemble techniques. Interpolation
kernel machines belong to the class of interpolating classifiers and do generalize well. They have been demonstrated to be a good alternative to
support vector machine for graph classification. In this work we further
improve their performance by considering multiple kernel learning. We
establish a general scheme for achieving this goal. The current experimental work is done using quadratic kernel combination. Our experimental
results demonstrate the performance boosting potential of our approach
against the use of individual graph kernels.

1 Introduction

There are a large amount of data that can be modeled naturally as graphs. They
appear in many application domains, ranging from social networks to chemistry
and medicine. Graph classification can be applied to such data for answering
numerous challenging questions [11,13,22]. For instance, chemical compounds
can be modeled as graphs where vertices correspond to atoms and edges to bonds.
Then, graph classification helps predict if a compound is active, say against HIV
or if a protein is an enzyme or not. The datasets used in our experimental work
as described in Sect. 4 give concrete examples of applications.

Graph kernels provide a way to compute similarities between graphs and have
been studied for a long time, e.g. for graph classification. Dozens of graph kernels
have been developed [11,13,22], which focus on specific structural properties of
graphs. The key design ideas behind these graph kernels differ, e.g. based on
Weisfeiler-Lehman test of graph isomorphism [24], paths [4], and graph decomposition [17]. Meanwhile, public graph kernel libraries are also available [10,25].

M. Vento et al. (Eds.): GbRPR 2023, LNCS 14121, pp. 3–14, 2023.
https://doi.org/10.1007/978-3-031-42795-4_1

Despite the diversity in the design of graph kernels, their use for classification is, however, rather monotonous and dominated by support vector machines (SVM). This is clearly reflected in the recent survey papers for graph kernels [11,13,22], e.g. "The criteria used for prediction are SVM for classification" [11]. This is not a surprise due to the dominance of SVM in machine learning.

As another kernel-based classification method, interpolation kernel machines belong to the class of interpolating classifiers that interpolate all the training data and thus have zero training error [27]. The study [3] demonstrated that interpolation kernel machines trained to have zero training error do perform very well on test data (a phenomenon typically observed in over-parametrized deep learning models). Our recent work [30] has shown that interpolation kernel machines are a good alternative to SVM for graph classification. This study justifies a systematic consideration of interpolation kernel machines parallel to the popular SVM for experimentation in graph classification. For this purpose we have proposed an extended experimental protocol in [30] to obtain the maximal possible graph classification performance in a particular application.

In this work we further improve the performance of interpolation kernel machines by considering multiple kernel learning [5]. We establish a general scheme for achieving this goal. The current experimental work is done using quadratic kernel combination. Our experimental results demonstrate the performance boosting potential of our approach.

The remainder of the paper is organized as follows. We give a brief general discussion of interpolating classifiers and introduce interpolation kernel machines in Sect. 2. The general scheme of multiple kernel learning tailored to interpolation kernel machines is presented in Sect. 3. The experimental results follow in Sect. 4. Finally, Sect. 5 concludes the paper.

2   Interpolating Classifiers

It is commonly believed that perfectly fitting the training data, as in the case of interpolating classifiers, must inevitably lead to overfitting. Recent research, however, reveals good reasons to study such classifiers. For instance, the work [27] provides strong indications that ensemble techniques are particularly successful if they are built on interpolating classifiers. A prominent example is random forest. Recently, Belkin [2] emphasizes the importance of interpolation (and its sibling over-parametrization) to understand the foundations of deep learning.

2.1   Interpolation Kernel Machines

Here we introduce a technique to fully interpolate the training data using kernel functions, known as kernel machines [3,9]. Note that this term has been often used in research papers (e.g. [8,29]), where variants of support vector machines are effectively meant. For the sake of clarity we will use the term "interpolation kernel machine" throughout the paper.

Let $X = \{x_1, x_2, \ldots, x_n\} \subset \Omega^n$ be a set of n training samples with their corresponding targets $Y = \{y_1, y_2, \ldots, y_n\} \subset T^n$ in the target space. The sets are sorted so that the corresponding training sample and target have the same index. A function $f : \Omega \to T$ interpolates this data iif:

$$f(x_i) = y_i, \quad \forall i \in 1, \ldots, n \tag{1}$$

Representer Theorem. Let $k : \Omega \times \Omega \to \mathbb{R}$ be a positive semidefinite kernel for some domain Ω, X and Y a set of training samples and targets as defined above, and $g : [0, \infty) \to \mathbb{R}$ a strictly monotonically increasing function for regularization. We define E as an error function that calculates the loss L of f on the whole sample set with:

$$E(X,Y) = E((x_1, y_1), \ldots, (x_n, y_n)) = \frac{1}{n} \sum_{i=1}^{n} L(f(x_i), y_i) + g(\|f\|) \tag{2}$$

Then, the function $f^* = \mathrm{argmin}_f \{E(X,Y)\}$ that minimizes the error E has the form:

$$f^*(z) = \sum_{i=1}^{n} \alpha_i k(z, x_i) \quad \text{with } \alpha_i \in \mathbb{R} \tag{3}$$

The proof can be found in many textbooks, e.g. [6].

We now can use f^* from Eq. (3) to interpolate our training data. Note that the only learnable parameters are $\alpha = (\alpha_1, \ldots, \alpha_n)$, a real-valued vector with the same length as the number of training samples. Learning α is equivalent to solving the system of linear equations:

$$G(\alpha_1^*, \ldots, \alpha_n^*)^T = (y_1, \ldots, y_n)^T \tag{4}$$

where $G \in \mathbb{R}^{n \times n}$ is the kernel (Gram) matrix. In case of positive definite kernel k the Gram matrix G is invertible. Therefore, we can find the optimal α^* to construct f^* by:

$$(\alpha_1^*, \ldots, \alpha_n^*)^T = G^{-1}(y_1, \ldots, y_n)^T \tag{5}$$

After learning, the interpolation kernel machine then uses the interpolating function from Eq. (3) to make prediction for test samples. Note that solving the optimal parameters α^* in (5) in a naive manner requires computation of order $\mathcal{O}(n^3)$ and is thus not feasible for large-scale applications. A highly efficient solver EigenPro has been developed [15] to enable significant speedup for training on GPUs. Another recent work [26] applies an explainable AI technique for sample condensation of interpolation kernel machines.

In this work we focus on classification problems. In this case $f(z)$ is encoded as a one-hot vector $f(z) = (f_1(z), \ldots f_c(z))$ with $c \in \mathbb{N}$ being the number of output classes. This requires c times repeating the learning process above, one for each component of the one-hot vector. This computation can be formulated as follows. Let $A_l = (\alpha_{l1}^*, ..., \alpha_{ln}^*)$ be the parameters to be learned and $Y_l = (y_{l1}, ..., y_{ln})$ target values for each component $l = 1, ..., c$. The learning of interpolation kernel machine becomes:

$$G \underbrace{\left(A_1^T, ..., A_c^T\right)}_{A} = \underbrace{\left(Y_1^T, ..., Y_c^T\right)}_{Y} \tag{6}$$

with the unique solution:

$$A = G^{-1} \cdot Y \tag{7}$$

which is the extended version of Eq. (5) for c classes and results in zero error on training data. When predicting a test sample z, the output vector $f(z)$ is not a probability vector in general. The class which gets the highest output value is considered as the predicted class. If needed, e.g. for the purpose of classifier combination, the output vector (z) can also be converted into a probability vector by applying the softmax function.

3   Multiple Kernel Learning for Interpolation Kernel Machines

Multiple kernel learning (MKL) methods use multiple kernels simultaneously instead of selecting one specific kernel function [5]. Given a pool of m kernels k_i, $i = 1, \ldots, m$, a combined kernel function is generally defined by:

$$\Phi(K = (k_1, \ldots, k_m), W)$$

where W are the weights (parameters) for kernel combination. Applications for MKL have been found, for instance, in processing biomedical data [16,23]. Recent developments include extensions to handle the scenario with missing channels in data [14] and unreliable data [1]. For graph classification, it was proposed to use a linear combination of kernel matrices obtained from the same kernel with different values of the hyper-parameters [18]. The recent work [28] deals with MKL using a GMDH-type neural network. In both [18,28] the classification backbone is SVM.

3.1   Construction of Combined Kernels

Positive definite kernels can be combined to build more complex positive definite kernels. Given two such kernels k_1 and k_2, some simple combination rules are:

- $k(x, y) = c \cdot k_1(x, y)$, for all $c \in \mathbb{R}_+$
- $k(x, y) = k_1(x, y) + c$, for all $c \in \mathbb{R}_+$
- $k(x, y) = k_1(x, y) + k_2(x, y)$
- $k(x, y) = k_1(x, y) \cdot k_2(x, y)$

– $k(x, y) = f(x) \cdot f(y)$ for any function f mapping to $\mathbb{R}$

The proof and additional combination rules can be found in [6]. The fundamental discussion in the following applies to any combined kernel $\Phi(K, W)$ that are constructed using such rules to guarantee the positive definiteness. Obviously, polynomial functions satisfy this requirement. This work will focus on quadratic combination only:

$$\Phi_q(K, W) = w_0 + \sum_{i=1}^{m} w_i \cdot k_i + \sum_{i=1}^{m} \sum_{j=i}^{m} w_{ij} \cdot k_i k_j$$

with the weights $W = (w_0, w_1, \ldots, w_m, w_{11}, w_{12}, \ldots, w_{1m}, \ldots, w_{mm})$, $|W| = (m+1)(m+2)/2$.

The weights W can be constrained in different ways. In this work we study three variants: $\mathbb{R}$ (no constraint), $\mathbb{R}_+^0$ (nonnegative), $[0, 1]$. When using nonnegative weights, we can extract and study the relative importance of the terms of combined kernels [5]. The interval $[0, 1]$ further restricts the influence of these terms.

3.2   General Scheme of MKL for Interpolation Kernel Machines

We present a general scheme (formulation and solution) of MKL tailored to interpolation kernel machines. Given the pool of kernels K, the combined kernel results in Gram matrix $G(W)$ that is built on $\Phi(K, W)$ and thus dependent of the weights W. Then, the task of learning an interpolation kernel machine becomes:

$$\min_{W, A} E(W, A) = \min_{W, A} \|G(W)A - Y\|^2 \qquad (8)$$

with $|W| + cn$ parameters to be optimally estimated.

Due to $|W| \ll cn$, this optimization problem is dominated by estimating the large number of parameters of the interpolation kernel machine. A parameter decomposition technique has been discussed in [12] for effective parameter reduction (see Appendix for a brief summary) and found applications [19, 32]. For applying this technique, we can naturally partition all parameters into two groups: W and A. Then, the optimization problem can be reformulated as:

$$\min_{W, A} E(W, A) = \min_{W} \left[\min_{A} \|G(W)A - Y\|^2 \right] \qquad (9)$$

$$= \min_{W} \Psi(W) = \min_{W} G(W)^{-1} \cdot Y \qquad (10)$$

where $\Psi(W)$ means the global minimum of the overall optimization function for a fixed W and all possibilities of A. The last step is justified by the fact that for a fixed W, we simply need to search for the optimal interpolation kernel machine for the concrete combined kernel $\Phi(K, W)$. When following the construction rules above, the combined kernel is guaranteed to remain positive definite. Thus, the corresponding Gram matrix $G(W)$ is invertible and the interpolation kernel machine deduced from $\Phi(K, W)$ has a unique solution, see Eq. (7). Doing it this way, we are able to reduce the number of optimization parameters from $|W| + cn$ to $|W|$ only without any loss of optimization quality.

In practice, however, the optimization scheme (10) does not work because of a specific property of interpolating classifiers, as an interpolation kernel machine is. The optimization function $E(W, A)$ is simply insensitive to W. Whatever value W is assumed to take, the interpolation kernel machine deduced from the combined kernel $\Phi(K, W)$ will produce zero training error, i.e. $\Psi(W) = 0$ holds for all W. Pragmatically, we thus introduce an approximation in order to make the optimization function $E(W, A)$ workable. For this purpose the original optimization task (8) is reformulated using the two-step (alternating) optimization method [5]:

– Step 1: We optimize A by using (fixing) the current W:

$$\min_A ||G(W)A - Y||^2$$

This corresponds to solving the standard problem of learning an interpolation kernel machine for the combined kernel $\Phi(K, W)$.
– Step 2: We optimize W by using (fixing) the current A:

$$\min_W ||G(W)A - Y||^2$$

– We repeat the two steps until convergence.

In both steps the minimization is done using the complete training dataset. This is exactly the reason for the trouble discussed above. Thus, we introduce a simple heuristic to circumvent the problem. We partition the training dataset T into two subsets: $T = T_1 \cup T_2$, $T_1 \cap T_2 = \emptyset$. Then, we use T_1 for step 1 and T_2 for step 2. In this work T_1 is chosen to contain 80% and T_2 20% of the training data, respectively.

3.3 Dealing with Indefinite Kernels

Learning interpolation kernel machines as presented in Sect. 2.1 requires positive definite kernels. In practice, however, many kernels reported in the literature, do not satisfy this property. This is particularly the case when working with graph kernels. Combining indefinite kernels generally does not deliver positive definite kernels. Thus, we need an indefinite interpolation kernel machine learning method for step 1 of the two-step optimization. Several methods for such an extension have been studied in [31]. We will use the matrix perturbation method in this work. The trouble with indefinite kernels is the lacking invertibility of the Gram matrix G. The matrix perturbation method artificially introduces some minor perturbation (noise) to G to make it invertible.

4   Experimental Results

We use 18 graph datasets from various domains (see Table 1 for an overview). These datasets have different application background. Examples of chemical datasets include benzodiazepine receptor dataset (BZR), cyclooxygenase-2 inhibitors (COX2), inhibitors of dihydrofolate reductase (DHFR), estrogen receptor ligands (ERMD). They are classified whether the chemical composition is active. Several datasets are related to medical applications. For instance, the AIDS dataset contains chemical compounds which have been screened as active or inactive against HIV. The Proteins dataset is for protein function prediction (functional class membership of enzymes and non-enzymes). The PTC (Predictive Toxicity Challenge) dataset is labeled according to carcinogenicity on rodents, aiming to predict carcinogenicity of chemical compounds. The Cuneiform dataset contains graphs representing different Hittite cuneiform signs. MSRC_9 and MSRC_21 are datasets in semantic image processing. Semantic labels are, for instance, building, grass, tree, cow, sky, sheep, boat, face, car, bicycle, etc.

Table 1. Description of graph datasets.

domain	dataset	# graphs	# classes	avg. # nodes	avg. # edges
Chemistry	BZR	405	2	35.8	38.4
	BZR_MD	306	2	21.3	225.1
	COX2	467	2	41.2	43.5
	COX2_MD	303	2	26.3	335.1
	DHFR	467	2	42.4	44.5
	DHFR_MD	393	2	23.87	283.01
	ER_MD	446	2	21.3	234.9
	MUTAG	188	2	17.9	19.8
Medicine	AIDS	2000	2	15.69	16.20
	PROTEINS	1113	2	39.1	72.8
	PROTEIN_full	1113	2	39.1	72.8
	PTC_FM	349	2	14.1	14.5
	PTC_FR	351	2	14.6	15.0
	PTC_MM	336	2	14.0	14.3
	PTC_MR	344	2	14.3	14.7
Vision	Cuneiform	267	30	21.3	44.8
	MSRC_9	221	8	40.6	97.9
	MSRC_21	563	20	77.5	198.3

Six graph kernels were selected for the experiments: shortest-path kernel [4], neighborhood hash kernel [7], Weisfeiler-Lehman graph kernel [24], tree-based graph decomposition kernel [17], propagation kernel [20], and pyramid match kernel [21]. The main selection criterion are the diverse design paradigms of these kernels. We used the implementations from the graph kernel library GraKeL written in Python [25]. For each pair of dataset and graph kernel, we conducted a 5-fold cross validation and report the average performance in term of classification accuracy.

The results on these datasets are presented in Table 2, where matrix perturbation was used for achieving positive definiteness. For each dataset the best-performing method is shown in bold. The quadratic kernel learning (QKL) considerably improves the performance (classification accuracy) of the individual graph kernels. In particular, restricting the weights to a small interval $[0, 1]$ turns out to be positive. The matrix perturbation method, however, is not best-performing among the methods studied in [31] to deal with indefinite kernels. Instead, a cross-entropy variant proved to be most favorable. Table 3 shows the superior performance of this variant compared with matrix perturbation on the individual graph kernels. Unfortunately, it is not possible to apply this variant in step 1 of the two-step optimization to replace matrix perturbation (the reason for this is rather technical and thus not detailed here). Nevertheless, we show a comparison with these results in Table 3. Here the quadratic kernel learning is still favorable.

Table 2. Accuracy (%) of individual graph kernels (matrix perturbation was used for achieving positive definiteness) and quadratic kernel learning.

dataset	method								
	Shortest Path	Neighborhood Hash	Weisfeiler Lehman	Graph Decomposition	Propagation	Pyramid Match	QKL ($\mathbb{R}$)	QKL ($\mathbb{R}_+^0$)	QKL ($[0,1]$)
BZR	70.1	**87.9**	83.9	85.7	70.4	67.9	84.9	86.4	86.7
BZR_MD	52.2	52.9	49.0	57.2	52.3	59.8	63.7	67.6	**68.6**
COX2	65.5	79.7	68.3	78.8	65.5	64.9	79.4	81.1	**81.2**
COX2_MD	60.7	48.2	52.5	53.8	53.4	55.8	60.7	63.0	**64.7**
DHFR	51.6	71.4	70.4	79.4	53.3	49.3	75.9	78.8	**79.9**
DHFR_MD	62.9	56.5	62.9	62.4	56.8	56.5	64.7	**68.7**	68.5
ER_MD	52.2	54.3	61.4	54.7	60.1	58.5	61.4	64.3	**69.7**
MUTAG	63.8	79.9	84.1	82.5	56.5	79.9	78.2	83.5	**84.6**
AIDS	63.5	94.6	81.7	52.5	49.3	66.2	99.1	**99.6**	99.5
PROTEINS	55.2	64.5	64.7	57.1	54.9	55.3	74.8	75.1	**75.5**
PROTEINS_full	57.3	63.4	64.8	55.6	58.9	57.8	76.4	**76.5**	76.5
PTC_FM	59.8	54.4	59.3	57.7	52.2	55.6	65.3	65.3	**66.2**
PTC_FR	60.4	61.8	62.7	62.7	52.1	61.0	63.3	**64.1**	64.1
PTC_MM	55.0	58.9	**64.3**	57.5	52.7	55.4	63.1	63.1	63.7
PTC_MR	54.4	57.9	54.7	59.0	52.9	57.3	62.2	61.0	**61.6**
Cuneiform	19.2	46.9	52.5	21.2	55.0	78.1	77.4	**82.1**	80.4
MSRC_9	51.5	92.8	90.1	85.9	90.5	92.3	95.5	**94.6**	94.2
MSRC_21	59.3	91.5	87.6	64.9	87.9	90.4	91.5	**91.8**	91.8
average	56.4	67.6	67.5	62.7	59.7	64.6	74.3	75.9	**76.5**

Table 3. Accuracy (%) of individual graph kernels (a cross-entropy variant was used to deal with indefinite kernels) and quadratic kernel learning.

dataset	method						
	Shortest Path	Neighborhood Hash	Weisfeiler Lehman	Graph Decomposition	Propagation	Pyramid Match	QKL ([0,1])
BZR	84.9	**87.8**	86.2	85.7	84.0	85.0	86.4
BZR_MD	68.9	66.7	58.5	62.7	57.8	59.8	**70.9**
COX2	79.7	79.7	81.0	78.8	80.1	80.5	**82.0**
COX2_MD	61.4	59.7	60.1	62.4	55.1	55.8	**66.3**
DHFR	75.7	**82.5**	79.6	79.4	75.9	77.4	79.7
DHFR_MD	67.2	60.4	64.4	67.9	60.5	56.5	**69.5**
ER_MD	62.8	69.3	67.7	54.7	67.7	58.5	**69.9**
MUTAG	78.8	84.6	79.8	82.5	74.5	**87.2**	84.6
AIDS	97.5	99.2	92.5	81.2	85.4	99.2	**99.5**
PROTEINS	74.9	73.7	68.5	65.4	68.9	73.9	**75.9**
PROTEINS_full	75.9	71.9	68.9	65.8	68.6	74.2	**76.5**
PTC_FM	60.7	58.5	**67.0**	62.2	61.6	59.9	65.6
PTC_FR	64.4	**68.4**	67.5	66.4	66.4	63.0	64.4
PTC_MM	63.7	67.9	67.0	66.1	**68.5**	63.7	62.8
PTC_MR	59.3	63.4	**64.2**	61.0	57.3	59.6	60.7
Cuneiform	70.7	71.7	81.5	75.4	**81.8**	78.1	80.4
MSRC_9	91.0	92.8	90.1	85.9	91.4	92.3	**95.2**
MSRC_21	59.3	91.5	87.6	64.9	82.4	90.4	**92.3**
average	72.0	75.0	74.0	70.5	71.6	73.1	**76.8**

5 Conclusion

Interpolation kernel machines have been demonstrated to be a good alternative to support vector machine for graph classification. In this work we further improve their performance by considering multiple kernel learning. We have presented a general scheme for achieving this goal. Our experimental results using quadratic kernel combination demonstrated the performance boosting potential of our approach against the use of individual graph kernels. In future we will study more complex combined kernels. In addition, we will also apply the proposed general scheme to non-graph data. With interpolation kernel machines as an alternative to support vector machine our work contributes to increasing the methodological plurality in the graph processing community (and beyond).

Acknowledgement. Jiaqi Zhang is supported by the China Scholarship Council (CSC). This research has received funding from the European Union's Horizon 2020 research and innovation programme under the Marie Sklodowska-Curie grant agreement No 778602 Ultracept.

Appendix: Parameter decomposition for function fitting

In contrast to treating all parameters simultaneously the parameters are decomposed into two parts, one part can be solved by either an analytic or a direct method, and the other part has to be solved by an iterative optimization process. Let $\mathbf{p}_i$, $i = 1, \ldots, n$, denote n data points to be fitted by function $f(\mathbf{v}) = 0$, where $\mathbf{v} \in \mathbb{R}^k$ is the parameter vector. Given the distance function $d(\mathbf{p}_i, f(\mathbf{v}))$ of a point $\mathbf{p}_i$ to $f(\mathbf{v})$, the geometric fitting problem can be formulated as:

$$\min_{\mathbf{v} \in \mathbb{R}^k} \sum_{i=1}^{n} d(\mathbf{p}_i, f(\mathbf{v}))$$

We decompose $\mathbf{v}$ into $\mathbf{v} = [\mathbf{v}_1 \mathbf{v}_2]$, where $\mathbf{v}_1 \in \mathbb{R}^{k_1}$, $\mathbf{v}_2 \in \mathbb{R}^{k_2}$, and $k \equiv k_1 + k_2$. After a rearrangement of parameters $\mathbf{v}$, the optimization task becomes:

$$\min_{\mathbf{v} \in \mathbb{R}^k} \sum_{i=1}^{n} d(\mathbf{p}_i, f(\mathbf{v})) = \min_{\mathbf{v}_1 \in \mathbb{R}^{k_1}} \left[\min_{\mathbf{v}_2 \in \mathbb{R}^{k_2}} \sum_{i=1}^{n} d(\mathbf{p}_i, f(\mathbf{v}_1, \mathbf{v}_2)) \right] = \min_{\mathbf{v}_1 \in \mathbb{R}^{k_1}} \Psi(\mathbf{v}_1)$$

where $\Psi(\mathbf{v}_1)$ means the global minimum sum of the distance measures for a fixed $\mathbf{v}_1$ and all possibilities of $\mathbf{v}_2$. Assume that the optimal $\mathbf{v}_2$ for reaching $\Psi(\mathbf{v}_1)$ can be solved analytically or in some other way, then the original optimization problem with a total of $k_1 + k_2$ parameters is transformed into an equivalent optimization problem with k_1 parameters only. This parameter reduction can be expected to decrease the number of local minima and thus reduce the possibilities of dropping into local minima. Moreover, this scheme also tends to reduce the computation time.

References

1. Alavi, F., Hashemi, S.: A bi-level formulation for multiple kernel learning via self-paced training. Pattern Recogn. **129**, 108770 (2022)
2. Belkin, M.: Fit without fear: remarkable mathematical phenomena of deep learning through the prism of interpolation. Acta Numerica **30**, 203–248 (2021)
3. Belkin, M., Ma, S., Mandal, S.: To understand deep learning we need to understand kernel learning. In: Proceedings of of 35th ICML, pp. 540–548 (2018)
4. Borgwardt, K.M., Kriegel, H.: Shortest-path kernels on graphs. In: Proceedings of 5th ICDM, pp. 74–81 (2005)
5. Gönen, M., Alpaydin, E.: Multiple kernel learning algorithms. J. Mach. Learn. Res. **12**, 2211–2268 (2011)
6. Herbrich, R.: Learning Kernel Classifiers: Theory and Algorithms. The MIT Press, Cambridge (2002)
7. Hido, S., Kashima, H.: A linear-time graph kernel. In: Proceedings of 9th ICDM, pp. 179–188 (2009)
8. Houthuys, L., Suykens, J.A.K.: Tensor-based restricted kernel machines for multi-view classification. Inf. Fusion **68**, 54–66 (2021)
9. Hui, L., Ma, S., Belkin, M.: Kernel machines beat deep neural networks on mask-based single-channel speech enhancement. In: Proceedings of 20th INTER-SPEECH, pp. 2748–2752 (2019)
10. Jia, L., Gaüzère, B., Honeine, P.: graphkit-learn: a python library for graph kernels based on linear patterns. Pattern Recogn. Lett. **143**, 113–121 (2021)
11. Jia, L., Gaüzère, B., Honeine, P.: Graph kernels based on linear patterns: theoretical and experimental comparisons. Expert Syst. Appl. **189**, 116095 (2022)
12. Jiang, X.: A decomposition approach to geometric fitting. In: Proceedings of IAPR Conference on Machine Vision Applications (MVA), pp. 467–470 (2000)

13. Kriege, N.M., Johansson, F.D., Morris, C.: A survey on graph kernels. Appl. Netw. Sci. **5**(1), 6 (2020)
14. Liu, X., et al.: Absent multiple kernel learning algorithms. IEEE Trans. Pattern Anal. Mach. Intell. **42**(6), 1303–1316 (2020)
15. Ma, S., Belkin, M.: Kernel machines that adapt to GPUs for effective large batch training. In: Proceedings of 3rd Conference on Machine Learning and Systems (2019)
16. Martínez-Vargas, J.D., Duque-Muñoz, L., Vargas-Bonilla, J.F., López, J.D., Castellanos-Domínguez, G.: Enhanced data covariance estimation using weighted combination of multiple Gaussian kernels for improved M/EEG source localization. Int. J. Neural Syst. **29**(6), 1950001:1–1950001:15 (2019)
17. Martino, G.D.S., Navarin, N., Sperduti, A.: A tree-based kernel for graphs. In: Proceedings of 12th SIAM International Conference on Data Mining, pp. 975–986 (2012)
18. Massimo, C.M., Navarin, N., Sperduti, A.: Hyper-parameter tuning for graph kernels via multiple kernel learning. In: Hirose, A., Ozawa, S., Doya, K., Ikeda, K., Lee, M., Liu, D. (eds.) ICONIP 2016. LNCS, vol. 9948, pp. 214–223. Springer, Cham (2016). https://doi.org/10.1007/978-3-319-46672-9_25
19. Mazzini, F., Kettler, D.T., Guerrero, J., Dubowsky, S.: Tactile robotic mapping of unknown surfaces, with application to oil wells. IEEE Trans. Instrument. Meas. **60**(2), 420–429 (2011)
20. Neumann, M., Garnett, R., Bauckhage, C., Kersting, K.: Propagation kernels: efficient graph kernels from propagated information. Mach. Learn. **102**(2), 209–245 (2016)
21. Nikolentzos, G., Meladianos, P., Vazirgiannis, M.: Matching node embeddings for graph similarity. In: Proceedings of 31st AAAI, pp. 2429–2435 (2017)
22. Nikolentzos, G., Siglidis, G., Vazirgiannis, M.: Graph kernels: a survey. J. Artif. Intell. Res. **72**, 943–1027 (2021)
23. Ruan, P., Hayashida, M., Akutsu, T., Vert, J.: Improving prediction of heterodimeric protein complexes using combination with pairwise kernel. BMC Bioinformatics **19**S(1), 73–84 (2018)
24. Shervashidze, N., Schweitzer, P., van Leeuwen, E.J., Mehlhorn, K., Borgwardt, K.M.: Weisfeiler-lehman graph kernels. J. Mach. Learn. Res. **12**, 2539–2561 (2011)
25. Siglidis, G., Nikolentzos, G., Limnios, S., Giatsidis, C., Skianis, K., Vazirgiannis, M.: GraKeL: a graph kernel library in python. J. Mach. Learn. Res. **21**, 54:1–54:5 (2020)
26. Winter, D., Bian, A., Jiang, X.: Layer-wise relevance propagation based sample condensation for kernel machines. In: Proceedings of 19th International Conference on Computer Analysis of Images and Patterns (CAIP), Part I, vol. 13052, pp. 487–496 (2021)
27. Wyner, A.J., Olson, M., Bleich, J., Mease, D.: Explaining the success of AdaBoost and random forests as interpolating classifiers. J. Mach. Learn. Res. **18**, 48:1–48:33 (2017)
28. Xu, L., et al.: Multiple graph kernel learning based on GMDH-type neural network. Inf. Fusion **66**, 100–110 (2021)
29. Xue, H., Chen, S.: Discriminality-driven regularization framework for indefinite kernel machine. Neurocomputing **133**, 209–221 (2014)

30. Zhang, J., Liu, C., Jiang, X.: Interpolation kernel machine and indefinite kernel methods for graph classification. In: Proceedings of 3rd International Conference on Pattern Recognition and Artificial Intelligence (ICPRAI). LNCS, vol. 13364, pp. 467–479. Springer, Heidelberg (2022). https://doi.org/10.1007/978-3-031-09282-4_39
31. Zhang, J., Liu, C.L., Jiang, X.: Indefinite interpolation kernel machines, submitted for publication (2023)
32. Zheng, G., Zhang, X.: A novel parameter decomposition based optimization approach for automatic pose estimation of distal locking holes from single calibrated fluoroscopic image. Pattern Recogn. Lett. **30**(9), 838–847 (2009)

Minimum Spanning Set Selection in Graph Kernels

Domenico Tortorella(✉) and Alessio Micheli

Department of Computer Science, University of Pisa, Largo B. Pontecorvo, 3,
56127 Pisa, Italy
domenico.tortorella@phd.unipi.it, micheli@di.unipi.it

Abstract. Kernel-based learning models such as support vector machines (SVMs) can seamlessly deal with graph structures thanks to suitable kernel functions that compute a particular similarity between pairs of data samples. In the last two decades, a plethora of graph kernels have been proposed, motivated by theoretical properties or designed specifically for an application domain. Computing graph kernels however presents a significant cost for both training and inference, since predictions on unseen graphs require evaluating the kernel e.g. between the new sample and all support vectors, and solutions to an SVM optimization problem are not guaranteed to be sparse. In this paper, we present a method to select a minimum set of spanning vectors for the solutions of SVMs, based on the rank-revealing QR decomposition of the kernel Gram matrix. We apply it on 18 real-world classification tasks on chemical compounds, showing its effectiveness to reduce the computational burden in performing inference on unseen graphs by minimizing the number of support vectors without penalizing accuracy. This in turn gives us a tool to better analyze the trade-off between accuracy, expressiveness and inference cost among different graph kernels.

Keywords: Graph Kernels · Support Vector Machines · Kernel basis · Rank-Revealing QR

1 Introduction

Among the structures commonly used to represent data, graphs are one of the most general, allowing to model complex objects as a set of entities (nodes) and relationships between such entities (edges) [10]. In chemo- and bio-informatics, in particular, graphs have been extensively used to represent molecular compounds [3,16,23], with nodes corresponding to atoms and edges corresponding to chemical bonds.

The flexibility afforded by graph structures has posed a challenge to traditional machine learning approaches conceived for vector or fixed-grid data. On one hand, artificial neural networks became able to adaptively learn graph representations by performing message-passing on the graph structure, leading

M. Vento et al. (Eds.): GbRPR 2023, LNCS 14121, pp. 15–24, 2023.
https://doi.org/10.1007/978-3-031-42795-4_2

to the current explosion of the graph neural network (GNN) models [1]. On the other hand, kernel-based learning models such as support vector machines (SVM) [18] were readily extended to graph structures by suitable kernel functions, which compute a particular similarity between pairs of data samples. In the last two decades a plethora of graph kernels have been proposed, tailored to a specific application domain or grounded in different theoretical properties of graphs [2,10].

Computing graph kernels presents a significant cost not only during training but also in the inference phase, since e.g. the decision function of an SVM requires evaluating the kernel between the new graph and all support vectors to make a prediction. Furthermore, the solutions of an SVM are not guaranteed to be sparse, leading to decision functions having a large proportion of the training set as support vectors [6]. Since this is a general issue that affects all kernel-based SVMs, different methods have been proposed to reduce the number of support vectors. Those approaches have mostly focused on the training phase often at the expense of accuracy, including clustering training samples [22], applying a divide-et-impera heuristic to select support vectors from smaller subsets of the training data [9], or finding a subset of samples sufficient for a separating surface between classes [11,15]. Even an altogether different type of SVM has been proposed in order to better control the number of support vectors [17]. All the aforementioned methods, however, do not offer guarantees against degradation of accuracy with respect to an SVM trained on the whole data.

The approach presented in this paper instead is applied after the training of an SVM to minimize the number of support vectors without any loss of accuracy. This method consists in selecting a spanning set of minimum cardinality able to represent all linear decisors in the kernel space that can be obtained from the data [6]. To the best of our knowledge, this is the first application of such method to graph kernels. The aim of our contribution is twofold: (i) to provide a simple and effective method to reduce the computational cost in performing inference on unseen graphs by minimizing the number of support vectors in SVMs without penalizing accuracy; (ii) to provide an approach to better analyze the trade-off between accuracy, expressivity and inference cost between different graph kernels.

The reminder of this paper is structured as follows. In Sect. 2 we present a brief background on SVMs and graph kernels. In Sect. 3 we introduce our method to select a spanning set for kernels based on the RRQR factorization. We validate this method experimentally in Sect. 4 on eighteen binary graph classification tasks, finally drawing conclusions in Sect. 5.

2   Support Vector Machines and Graph Kernels

Let $\mathcal{X}$ be a set of n data samples $\{x_i\}_{i=1}^n$ divided into two classes identified by a label $y_i \in \{\pm 1\}$ associated to each one. A binary classifier is a function $\hat{y} : \mathcal{X} \to \{\pm 1\}$ that tries to predict the correct class y_i of a sample $x_i \in \mathcal{X}$.

A kernel function $\kappa : \mathcal{X} \times \mathcal{X} \to \mathbb{R}$ is a symmetric positive semi-definite function that represents a scalar product in a reproducing kernel Hilbert space

(RKHS). This function allows to work in the RKHS without having to explicitly define the transformation from $\mathcal{X}$ into such space. For example, we can define a hyper-plane to separate the samples belonging to the two classes as the locus $f(\boldsymbol{x}) = 0$ of the function parametrized by $\boldsymbol{\vartheta} \in \mathbb{R}^n, \beta \in \mathbb{R}$:

$$f(\boldsymbol{x}) = \sum_{i=1}^{n} \vartheta_i \kappa(\boldsymbol{x}_i, \boldsymbol{x}) + \beta. \tag{1}$$

A kernel-based support vector machine [18] learns a classifier $\hat{y}(\boldsymbol{x}) = \text{sign}(f(\boldsymbol{x}))$ by finding the maximum-margin separating hyper-plane as the solution of the following optimization problem:

$$\min_{\boldsymbol{\vartheta} \in \mathbb{R}^n, \beta \in \mathbb{R}} \frac{1}{2} \sum_{i=1}^{n} \sum_{j=1}^{n} \vartheta_i \vartheta_j \kappa(\boldsymbol{x}_i, \boldsymbol{x}_j) + C \sum_{i=1}^{n} \max\{0, 1 - y_i f(\boldsymbol{x}_i)\}, \tag{2}$$

with $C > 0$ a constant affecting the trade-off between model regularization and empirical error, where a larger value of C favors solutions of (2) with smaller misclassification margins over smaller norms of $\boldsymbol{\vartheta}$. The samples $\boldsymbol{x}_i$ that correspond to coefficients $\vartheta_i \neq 0$ are defined as the support vectors (SVs) of the SVM.

2.1 Graph Kernels

Note that so far we have not specified the nature of data samples in $\mathcal{X}$. Indeed, kernel-based learning allows to deal with any kind of data structure beyond vectors, such as sequences, trees, and graphs. In the latter case, the data samples $\boldsymbol{x}_i \in \mathcal{X}$ are objects $\boldsymbol{x}_i = (\mathcal{V}_i, \mathcal{E}_i, \ell_i)$, where $\mathcal{V}_i$ is the set of nodes, $\mathcal{E}_i \subseteq \mathcal{V}_i \times \mathcal{V}_i$ is the set of edges, and $\ell_i(v) \in \mathcal{L}$ is a label that can be associated with each node $v \in \mathcal{V}_i$. A plethora of kernel functions for graphs has been proposed so far, based on different theoretical features of graphs or motivated by a specific application domain [2, 10]. Generally speaking, a graph kernel function computes a similarity measure between a pair of graphs based on common properties or sub-structures. Some examples of graph kernels include:

- Graphlet (GL) kernel [20], which compares the distribution of all possible small sub-graphs (so-called "graphlets") having up to k nodes;
- Shortest-path (SP) kernel [4], which decomposes graphs into shortest paths and compares pairs of shortest paths according to their lengths and the labels of their endpoints;
- Weisfeiler–Lehman (WL) kernel [19], which is based on h iterations of the WL graph isomorphism test and is equivalent to comparing the number of shared sub-trees of height h between the two graphs.

Computing a graph kernel can be significantly costly. For example, the SP kernel must compare all possible pairs of shortest paths from the two graphs, while the WL kernel scales quadratically in the total number of graphs (support vectors plus the graph to infer) due to the multi-set relabeling step of the WL test iteration. Therefore, keeping at minimum the number of support vectors in the decision function (1) is of paramount concern for the efficiency of SVMs and other kernel-based learners.

3   Kernel Spanning Set Selection via RRQR

Let us focus on the linear part of the function $f(\boldsymbol{x})$ defined in (1). This is an element belonging to the dual space of $\mathcal{X}$ with respect to the kernel κ, that is the vector space of functions $\boldsymbol{z} \mapsto \kappa(\boldsymbol{x}_i, \boldsymbol{z})$ and their linear combinations

$$\mathcal{F} = \mathrm{Span}\{\kappa(\boldsymbol{x}_1, \cdot), ..., \kappa(\boldsymbol{x}_n, \cdot)\} \stackrel{\text{def}}{=} \left\{ \sum_{i=1}^{n} \vartheta_i \kappa(\boldsymbol{x}_i, \cdot) : \boldsymbol{\vartheta} \in \mathbb{R}^n \right\}. \tag{3}$$

Any subset $\mathcal{S} \subseteq \mathcal{X}$ such that $\mathrm{Span}\{\kappa(\boldsymbol{x}, \cdot) : \boldsymbol{x} \in \mathcal{S}\} = \mathcal{F}$ is called a spanning set of $\mathcal{F}$. The smallest cardinality of a spanning set is the dimension of the vector space $\mathcal{F}$, and such spanning sets are called basis of $\mathcal{F}$. We define the rank r of kernel κ on $\mathcal{X}$ as the dimension of $\mathcal{F}$. To reduce the number of support vectors from n to n', we must find a basis of the subspace of $\mathcal{F}$ spanned by the SVs, along with new coefficients $\boldsymbol{\vartheta}' \in \mathbb{R}^{n'}$ so that

$$f(\boldsymbol{x}) = \sum_{i=1}^{n} \vartheta_i \kappa(\boldsymbol{x}_i, \boldsymbol{x}) + \beta = \sum_{i=1}^{n'} \vartheta'_i \kappa(\boldsymbol{x}_{s_i}, \boldsymbol{x}) + \beta = f'(\boldsymbol{x}). \tag{4}$$

To obtain these, we work on the Gram matrix $\mathbf{K}$, whose elements are $K_{ij} = \kappa(\boldsymbol{x}_i, \boldsymbol{x}_j)$, and is thus a real symmetric matrix. Each of its columns can be seen as a representation in vector form of $\kappa(\boldsymbol{x}_i, \cdot)$, since its evaluation on all elements of $\mathcal{X}$ completely specifies it. Therefore the linear combinations that span the range of $\mathbf{K}$ have the same coefficients of the linear combinations of vectors in $\mathcal{F}$. Let us illustrate this with a simple example.

Example. Consider a set of $n = 4$ samples $\mathcal{X} = \{\boldsymbol{x}_1, \boldsymbol{x}_2, \boldsymbol{x}_3, \boldsymbol{x}_4\}$ and a kernel function κ defined between pairs of its elements such that its Gram matrix is

$$\mathbf{K} = \begin{pmatrix} \kappa(\boldsymbol{x}_1, \boldsymbol{x}_1) & \kappa(\boldsymbol{x}_1, \boldsymbol{x}_2) & \kappa(\boldsymbol{x}_1, \boldsymbol{x}_3) & \kappa(\boldsymbol{x}_1, \boldsymbol{x}_4) \\ \kappa(\boldsymbol{x}_2, \boldsymbol{x}_1) & \kappa(\boldsymbol{x}_2, \boldsymbol{x}_2) & \kappa(\boldsymbol{x}_2, \boldsymbol{x}_3) & \kappa(\boldsymbol{x}_2, \boldsymbol{x}_4) \\ \kappa(\boldsymbol{x}_3, \boldsymbol{x}_1) & \kappa(\boldsymbol{x}_3, \boldsymbol{x}_2) & \kappa(\boldsymbol{x}_3, \boldsymbol{x}_3) & \kappa(\boldsymbol{x}_3, \boldsymbol{x}_4) \\ \kappa(\boldsymbol{x}_4, \boldsymbol{x}_1) & \kappa(\boldsymbol{x}_4, \boldsymbol{x}_2) & \kappa(\boldsymbol{x}_4, \boldsymbol{x}_3) & \kappa(\boldsymbol{x}_4, \boldsymbol{x}_4) \end{pmatrix} = \begin{pmatrix} 1 & 1 & 2 & 2 \\ 1 & 2 & 3 & 2 \\ 2 & 3 & 5 & 4 \\ 2 & 2 & 4 & 4 \end{pmatrix}.$$

Each column (or row) of $\mathbf{K}$ represents a function $\kappa(\boldsymbol{x}_i, \cdot)$ by its evaluation on all samples in $\mathcal{X}$. For example, $\kappa(\boldsymbol{x}_1, \cdot)$ is represented as $(1\ 1\ 2\ 2)^\top$. In this way is easy to notice that $\kappa(\boldsymbol{x}_3, \cdot) = \kappa(\boldsymbol{x}_1, \cdot) + \kappa(\boldsymbol{x}_2, \cdot)$ and $\kappa(\boldsymbol{x}_4, \cdot) = 2\kappa(\boldsymbol{x}_1, \cdot)$. Since $\mathrm{rank}(\mathbf{K}) = 2$, this means that the first two columns are linearly independent and provide a minimum spanning set. Therefore, in our example we have that $\mathcal{F} = \mathrm{Span}\{\kappa(\boldsymbol{x}_1, \cdot), \kappa(\boldsymbol{x}_2, \cdot)\}$, and that any $f(\boldsymbol{x})$ of (1) can be expressed as

$$f'(\boldsymbol{x}) = (\vartheta_1 + \vartheta_3 + 2\vartheta_4)\,\kappa(\boldsymbol{x}_1, \boldsymbol{x}) + (\vartheta_2 + \vartheta_3)\,\kappa(\boldsymbol{x}_2, \boldsymbol{x}) + \beta.$$

For an SVM decision function, that would mean halving the number of support vectors without affecting the prediction score. $\square$

To find a set of spanning columns for the Gram matrix $\mathbf{K}$, we rely on a rank-revealing QR (RRQR) decomposition [5,8]. This matrix factorization consists of permuting the columns of $\mathbf{K}$ such that the resulting QR factorization contains an upper triangular matrix whose linearly independent columns are separated from the linearly dependent ones. In detail, the RRQR factorization produces a column permutation $\mathbf{\Pi}$, an orthogonal matrix $\mathbf{Q}$, and an upper triangular matrix $\mathbf{R}$ such that

$$
\mathbf{\Pi K} = \left(\begin{array}{c|c} \bar{\mathbf{K}}_{rr} & \bar{\mathbf{K}}_{rn} \\ \hline \bar{\mathbf{K}}_{nr} & \bar{\mathbf{K}}_{nn} \end{array} \right) = \mathbf{Q} \left(\begin{array}{ccc|ccc} R_{1,1} & \cdots & R_{1,r} & R_{1,r+1} & \cdots & R_{1,n} \\ \vdots & \ddots & \vdots & \vdots & & \vdots \\ 0 & \cdots & R_{r,r} & R_{r,r+1} & \cdots & R_{r,n} \\ \hline 0 & \cdots & 0 & R_{r+1,r+1} & \cdots & R_{r+1,n} \\ \vdots & & \vdots & & \ddots & \vdots \\ 0 & \cdots & 0 & 0 & \cdots & R_{n,n} \end{array} \right). \tag{5}
$$

The singular values of $\mathbf{R}$ are obtained from the diagonal elements as $|R_{i,i}|$. In (5), the singular values $|R_{1,1}|, ..., |R_{r,r}| > 0$ correspond to the r columns of $\mathbf{\Pi K}$ that are linearly independent, while the $|R_{r+1,r+1}|, ..., |R_{n,n}| \approx 0$ correspond to the linearly dependent columns. The computational cost of the whole QR decomposition is $O(n^3)$, but it can be reduced to $O(rn^2)$ by halting as soon as $|R_{i,i}| \approx 0$. A strong RRQR [7,14] additionally improves the numerical stability of the factorization by ensuring that certain bounds on the singular values of $\mathbf{R}$ are respected.

At this point we have discovered both the rank r of our kernel κ as the number of nonzero singular values of $\mathbf{R}$, and a set of samples $\mathcal{S} = \{\boldsymbol{x}_{\pi^{-1}(1)}, ..., \boldsymbol{x}_{\pi^{-1}(r)}\}$ that span the dual space of $\mathcal{X}$ (here $\pi(\cdot)$ is the permutation represented by the matrix $\mathbf{\Pi}$). We can finally make explicit the function $f'(\boldsymbol{x})$ of (4) as

$$
f'(\boldsymbol{x}) = \sum_{i=1}^{r} \vartheta_i' \kappa(\boldsymbol{x}_{\pi^{-1}(i)}, \boldsymbol{x}) + \beta, \tag{6}
$$

where the coefficient vector $\boldsymbol{\vartheta}' \in \mathbb{R}^r$ is computed from $\boldsymbol{\vartheta} \in \mathbb{R}^n$ as

$$
\boldsymbol{\vartheta}' = \left(\mathbf{I}_r \mid \bar{\mathbf{K}}_{rr}^{-1} \bar{\mathbf{K}}_{rn} \right) \mathbf{\Pi}\, \boldsymbol{\vartheta}. \tag{7}
$$

4 Experiments and Discussion

We now investigate experimentally the kernel rank and the minimum spanning set selection of support vectors on eighteen real-word graph datasets, representing binary classification tasks of chemical compounds such as toxicity or carcinogenicity [12]. We focus our analysis on the GL kernel with graphlet size $k = 5$, the shortest-path kernel (SP), and the WL kernel with $h = 4$ iterations. The kernels have been computed with the GraKeL library [21]. The number of data samples and the ranks of the three graph kernels are reported in Table 1.

Table 1. Ranks of three graph kernels for binary classification datasets, along with the SVM cross-validation accuracy on 10 folds.

Dataset	Samples	Kernel rank			SVM accuracy		
		GL	SP	WL	GL	SP	WL
AIDS	2000	9	538	1866	$80.0_{\pm0.0}$	$99.7_{\pm0.4}$	$98.4_{\pm0.8}$
BZR	405	7	264	390	$78.8_{\pm1.1}$	$87.9_{\pm3.2}$	$89.4_{\pm2.8}$
BZR_MD	306	1	24	202	$51.3_{\pm1.0}$	$73.2_{\pm5.7}$	$59.5_{\pm6.0}$
COX2	467	8	312	373	$78.2_{\pm0.9}$	$83.7_{\pm3.2}$	$83.7_{\pm2.4}$
COX2_MD	303	1	26	228	$51.2_{\pm1.4}$	$68.3_{\pm11.7}$	$58.8_{\pm3.9}$
DD	1178	21	1178	1178	$74.5_{\pm3.7}$	$80.6_{\pm2.2}$	$79.5_{\pm1.7}$
DHFR	756	5	327	704	$61.0_{\pm0.5}$	$81.7_{\pm2.6}$	$85.5_{\pm4.4}$
DHFR_MD	393	1	26	244	$67.9_{\pm1.2}$	$70.2_{\pm3.0}$	$69.0_{\pm1.8}$
ER_MD	446	1	29	240	$59.4_{\pm0.7}$	$68.2_{\pm5.7}$	$68.4_{\pm6.4}$
Mutagenicity	4337	8	635	4147	$67.3_{\pm2.4}$	$77.9_{\pm2.2}$	$84.3_{\pm1.7}$
MUTAG	188	4	78	174	$71.8_{\pm6.1}$	$88.2_{\pm8.7}$	$88.2_{\pm6.1}$
NCI1	4110	9	766	3915	$59.9_{\pm2.2}$	$74.2_{\pm2.1}$	$85.0_{\pm1.7}$
NCI109	4127	8	734	3931	$60.4_{\pm1.9}$	$73.8_{\pm1.7}$	$85.1_{\pm1.7}$
PTC_FM	349	7	205	331	$63.3_{\pm4.4}$	$66.8_{\pm5.4}$	$68.5_{\pm3.6}$
PTC_FR	351	7	207	335	$66.1_{\pm0.9}$	$72.9_{\pm3.3}$	$70.1_{\pm2.8}$
PTC_MM	336	7	202	317	$64.0_{\pm2.6}$	$68.4_{\pm5.1}$	$69.0_{\pm4.5}$
PTC_MR	344	7	204	328	$59.6_{\pm3.8}$	$66.5_{\pm6.2}$	$66.6_{\pm6.0}$
PROTEINS	1113	21	178	1067	$69.7_{\pm3.0}$	$77.7_{\pm3.3}$	$76.9_{\pm4.0}$

We can immediately notice how the basis for the ranks of GL kernels are considerably smaller compared to the number of samples and the corresponding ranks of the SP and WL kernels.

We split the datasets in 10 balanced folds, reserving 90% of the samples for training the SVM and 10% for selecting the regularization parameter C in the range $\{10^{-5}, ..., 10^4\}$. We report the average validation accuracy in Table 1 and the average number of support vectors in Table 2. In our evaluation we take a model selection perspective, with the aim of choosing the least computationally-demanding model without penalizing accuracy and of analyzing the trade-offs between the number of support vectors and model accuracy.

We apply the methods of Sect. 3 to select a minimum spanning set and reduce the number of support vectors for each selected SVM. Comparing the average number of support vectors and spanning vectors reported in Table 2, we notice that the SVs are often a large portion of the samples for all three kernels, while the required spanning vectors are considerably less. We can thus achieve a reduction of up to 99% in most of the cases for the GL kernel and of at least 10% for the SP kernel in most of the cases, while for the WL kernel this is generally limited below 5%. Since we have found no difference between the predictions of the original and the reduced SVMs, instead of the validation accuracy of the latter we report the maximum approximation error of the score function, defined as $||f' - f||_\infty = \max_{x \in \mathcal{X}} |f'(x) - f(x)|$. In most of the cases this is smaller than 10^{-5}, thanks to the numerical stability of the RRQR decomposition.

Table 2. Average number of support vectors (SVs), spanning set vectors, and average relative reduction on 10-fold cross-validation, along with the maximum approximation error of the SVM scoring function. (Average SVs and spanning set vectors are rounded.)

Dataset	SVs			Spanning			Reduction			Approximation		
	GL	SP	WL	GL	SP	WL	GL	SP	WL	GL	SP	WL
AIDS	720	119	456	8	107	452	−98.9%	−10.3%	−0.9%	$10^{-5.8}$	$10^{-8.0}$	$10^{-9.2}$
BZR	155	140	191	4	140	191	−97.4%	−0.1%	0%	$10^{-7.3}$	$10^{-7.8}$	10^{-10}
BZR_MD	268	203	254	1	23	188	−99.6%	−88.5%	−26.1%	$10^{-8.2}$	$10^{-7.7}$	$10^{-8.9}$
COX2	184	170	230	4	168	227	−97.8%	−0.9%	−1.5%	$10^{-7.3}$	$10^{-7.7}$	$10^{-9.3}$
COX2_MD	266	222	256	1	25	210	−99.6%	−88.7%	−17.9%	$10^{-7.6}$	$10^{-7.3}$	$10^{-8.8}$
DD	653	591	892	21	591	892	−96.7%	0%	0%	$10^{-5.5}$	$10^{-8.7}$	$10^{-9.1}$
DHFR	531	316	361	4	241	352	−99.2%	−23.9%	−2.3%	$10^{-6.4}$	$10^{-2.2}$	$10^{-8.8}$
DHFR_MD	227	229	251	1	24	168	−99.5%	−89.4%	−33.0%	$10^{-7.5}$	$10^{-7.2}$	$10^{-8.6}$
ER_MD	326	299	332	1	21	212	−99.6%	−93.0%	−36.1%	10^{-12}	$10^{-6.8}$	$10^{-8.5}$
Mutagenicity	2868	2062	2155	8	508	2143	−99.7%	−75.3%	−0.5%	$10^{-5.5}$	$10^{-5.1}$	$10^{-7.1}$
MUTAG	113	64	97	4	52	96	−96.5%	−17.9%	−0.3%	$10^{-8.1}$	$10^{-8.1}$	10^{-11}
NCI1	3356	2311	2253	8	610	2241	−99.7%	−73.5%	−0.5%	$10^{-3.7}$	$10^{-0.7}$	$10^{-6.9}$
NCI109	3281	2434	2301	8	600	2290	−99.7%	−75.3%	−0.4%	$10^{-3.7}$	$10^{-2.8}$	$10^{-6.7}$
PTC_FM	254	245	284	7	161	276	−97.2%	−34.3%	−2.8%	$10^{-7.5}$	$10^{-6.9}$	$10^{-8.6}$
PTC_FR	219	183	270	7	141	265	−96.8%	−22.9%	−1.8%	$10^{-7.5}$	$10^{-6.8}$	$10^{-9.0}$
PTC_MM	233	218	261	7	145	251	−96.9%	−33.4%	−3.9%	$10^{-7.8}$	$10^{-7.3}$	$10^{-9.2}$
PTC_MR	269	234	284	7	155	277	−97.3%	−33.9%	−2.4%	$10^{-7.5}$	$10^{-6.9}$	$10^{-8.8}$
PROTEINS	664	588	777	21	139	773	−96.8%	−76.3%	−0.5%	$10^{-5.6}$	$10^{-2.4}$	$10^{-8.1}$

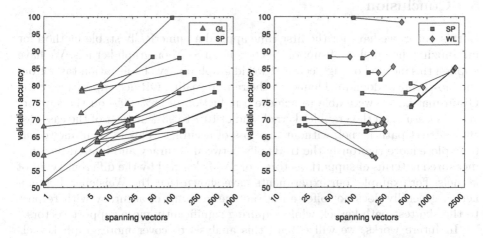

Fig. 1. Trade-off between spanning set size and accuracy for the graph kernels considered in our experiments. Lines join the pairs of results belonging to the same task.

Having both determined the accuracy and the actual number of required support vectors, we are now able to analyze the trade-offs offered by the three kernels we have considered in our experiments. In Fig. 1 we can observe that the SP kernel requires up to two orders of magnitude more support vectors with respect to the GL kernel, with a gain in terms of accuracy from 5% up to 20%. The WL kernel, on the other hand, while requiring a larger number of support vectors, it does not offer in exchange a gain in accuracy compared to the SP kernel in almost all of the cases. This analysis based essentially on the rank of graph kernels can be seen as complementary to that of [13], which investigates the expressiveness of graph kernels from a statistical learning theory perspective.

We point out that the significant reduction in support vectors could allow more insights into how the learned model takes its decisions. For example, from a spanning set of the GL kernel on the MUTAG dataset reported in Fig. 2 an expert could readily analyze the molecular functional groups that correspond to relevant graphlets used in computing the graph similarity by the kernel. We will explore the repercussions of our work on model explainability in future research.

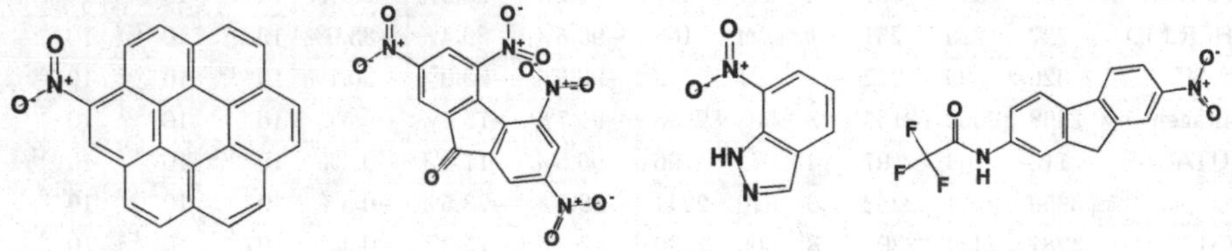

Fig. 2. A minimum spanning set for the GL kernel on the MUTAG chemical dataset.

5   Conclusion

In this paper, we have for the first time applied a numerically stable method for minimizing the number of support vectors in an SVM on graph kernels. We have applied this method on eighteen real-world graph binary classification tasks from the bio-chemical domain. Thanks to the rank-revealing QR decomposition of the Gram matrix, we were able to achieve a reduction of 96%–99% on the support vectors learned on the graphlet kernel, and of 10% to 93% in most of the cases for the shortest-path kernel, without any loss of accuracy. This in turn allowed us to explore more rigorously the trade-offs between accuracy and inference cost—measured in terms of support vectors—of SVMs learned by the different types of kernels. For example, our experiments have shown that the Weisfeiler–Lehman kernel generally does not offer an advantage in terms of accuracy with respect to the shortest-path kernel, while requiring significantly more support vectors.

In future works, we will extend this analysis to cover more graph kernels and the effect of their parameters, such as the graphlet size or the number of

Weisfeiler–Lehman isomorphism test iterations. This will offer a perspective complementary to the previous analysis of the expressiveness of graph kernels based on statistical learning theory. We will also investigate the accuracy/efficiency trade-offs posed by lower-rank approximations, that is discarding also linearly-independent support vectors, and explore the possibility of directly training SVMs on a minimum spanning subset of samples.

Acknowledgement. Research partly funded by PNRR - M4C2 - Investimento 1.3, Partenariato Esteso PE00000013 - "FAIR - Future Artificial Intelligence Research" - Spoke 1 "Human-centered AI", funded by the European Commission under the NextGeneration EU programme.

References

1. Bacciu, D., Errica, F., Micheli, A., Podda, M.: A gentle introduction to deep learning for graphs. Neural Netw. **129**, 203–221 (2020). https://doi.org/10.1016/j.neunet.2020.06.006
2. Borgwardt, K., Ghisu, E., Llinares-López, F., O'Bray, L., Rieck, B.: Graph kernels: state-of-the-art and future challenges. Found. Trends Mach. Learn. **13**(5–6), 531–712 (2020). https://doi.org/10.1561/2200000076
3. Borgwardt, K.M., Ong, C.S., Schonauer, S., Vishwanathan, S.V.N., Smola, A.J., Kriegel, H.P.: Protein function prediction via graph kernels. Bioinformatics **21**(Suppl. 1), i47–i56 (2005). https://doi.org/10.1093/bioinformatics/bti1007
4. Borgwardt, K., Kriegel, H.: Shortest-path kernels on graphs. In: Proceedings of the Fifth IEEE International Conference on Data Mining, pp. 74–81 (2005). https://doi.org/10.1109/ICDM.2005.132
5. Chandrasekaran, S., Ipsen, I.C.F.: On rank-revealing factorisations. SIAM J. Matrix Anal. Appl. **15**(2), 592–622 (1994). https://doi.org/10.1137/S0895479891223781
6. Downs, T., Gates, K., Masters, A.: Exact simplification of support vector solutions. J. Mach. Learn. Res. **2**(Dec), 293–297 (2001). https://www.jmlr.org/papers/v2/downs01a.html
7. Gu, M., Eisenstat, S.C.: Efficient algorithms for computing a strong rank-revealing QR factorization. SIAM J. Sci. Comput. **17**(4), 848–869 (1996). https://doi.org/10.1137/0917055
8. Hong, Y.P., Pan, C.T.: Rank-revealing QR factorizations and the singular value decomposition. Math. Comput. **58**(197), 213–232 (1992). https://doi.org/10.2307/2153029
9. Joachims, T.: Making large-scale support vector machine learning practical. In: Schölkopf, B., Burges, C.J.C., Smola, A.J. (eds.) Advances in Kernel Methods: Support Vector Learning, Chapter 11, pp. 169–184 (1999). https://doi.org/10.7551/mitpress/1130.003.0015
10. Kriege, N.M., Johansson, F.D., Morris, C.: A survey on graph kernels. Appl. Netw. Sci. **5**(1), 1–42 (2019). https://doi.org/10.1007/s41109-019-0195-3
11. Lee, Y.J., Mangasarian, O.L.: RSVM: reduced support vector machines. In: Proceedings of the 2001 SIAM International Conference on Data Mining. Society for Industrial and Applied Mathematics, Philadelphia, PA (2001). https://doi.org/10.1137/1.9781611972719.13

12. Morris, C., Kriege, N.M., Bause, F., Kersting, K., Mutzel, P., Neumann, M.: TUDataset: a collection of benchmark datasets for learning with graphs. In: ICML 2020 Workshop on Graph Representation Learning and Beyond (GRL+ 2020) (2020). https://www.graphlearning.io
13. Oneto, L., Navarin, N., Donini, M., Sperduti, A., Aiolli, F., Anguita, D.: Measuring the expressivity of graph kernels through statistical learning theory. Neurocomputing **268**, 4–16 (2017). https://doi.org/10.1016/j.neucom.2017.02.088
14. Pan, C.T., Tang, P.T.P.: Bounds on singular values revealed by QR factorizations. BIT Numer. Math. **39**(4), 740–756 (1999). https://doi.org/10.1023/A:1022395308695
15. Panja, R., Pal, N.R.: MS-SVM: minimally spanned support vector machine. Appl. Soft Comput. J. **64**, 356–365 (2018). https://doi.org/10.1016/j.asoc.2017.12.017
16. Ralaivola, L., Swamidass, S.J., Saigo, H., Baldi, P.: Graph kernels for chemical informatics. Neural Netw. **18**(8), 1093–1110 (2005). https://doi.org/10.1016/j.neunet.2005.07.009
17. Schölkopf, B., Smola, A.J., Williamson, R.C., Bartlett, P.L.: New support vector algorithms. Neural Comput. **12**(5), 1207–1245 (2000). https://doi.org/10.1162/089976600300015565
18. Schölkopf, B., Smola, A.J.: Learning with Kernels: Support Vector Machines, Regularization, Optimization, and Beyond. The MIT Press, Cambridge (2001). https://doi.org/10.7551/mitpress/4175.001.0001
19. Shervashidze, N., Schweitzer, P., van Leeuwen, E.J., Mehlhorn, K., Borgwardt, K.M.: Weisfeiler-Lehman graph kernels. J. Mach. Learn. Res. **12**, 2539–2561 (2011). https://doi.org/10.5555/1953048.2078187
20. Shervashidze, N., Vishwanathan, S.V., Petri, T.H., Mehlhorn, K., Borgwardt, K.M.: Efficient graphlet kernels for large graph comparison. In: Proceedings of the Twelth International Conference on Artificial Intelligence and Statistics, vol. 5, pp. 488–495 (2009). https://proceedings.mlr.press/v5/shervashidze09a.html
21. Siglidis, G., Nikolentzos, G., Limnios, S., Giatsidis, C., Skianis, K., Vazirgiannis, M.: GraKeL: a graph kernel library in Python. J. Mach. Learn. Res. **21**(54), 1–5 (2020). http://jmlr.org/papers/v21/18-370.html
22. Tran, Q.A., Zhang, Q.L., Li, X.: Reduce the number of support vectors by using clustering techniques. In: Proceedings of the Second International Conference on Machine Learning and Cybernetics, vol. 2, pp. 1245–1248 (2003). https://doi.org/10.1109/icmlc.2003.1259678
23. Yi, H.C., You, Z.H., Huang, D.S., Kwoh, C.K.: Graph representation learning in bioinformatics: trends, methods and applications. Brief. Bioinform. **23**(1), 1–16 (2022). https://doi.org/10.1093/bib/bbab340

Graph-Based vs. Vector-Based Classification: A Fair Comparison

Anthony Gillioz[1]([⊠]) [iD] and Kaspar Riesen[1,2] [iD]

[1] Institute of Computer Science, University of Bern, Bern, Switzerland
{anthony.gillioz,kaspar.riesen}@unibe.ch
[2] Institute for Information Systems, University of Applied Science and Arts
Northwestern Switzerland, Olten, Switzerland

Abstract. Numerous graph classifiers are readily available and frequently used in both research and industry. Ensuring their performance across multiple domains and applications is crucial. In this paper, we conduct a comprehensive assessment of three commonly used graph-based classifiers across 24 graph datasets (we employ classifiers based on graph matchings, graph kernels, and graph neural networks). Our goal is to find out what primarily affects the performance of these classifiers in different tasks. To this end, we compare each of the three classifiers in three different scenarios. In the first scenario, the classifier has access to the original graphs, in the second scenario, the same classifier has access only to the structure of the graph (without labels), and in the third scenario, we replace the graph-based classifiers with a corresponding related statistical classifier, which has access only to an aggregated feature vector of the graph labels. On the basis of this exhaustive evaluation, we are able to suggest whether or not certain graph datasets are suitable for specific benchmark comparisons.

Keywords: Graph Classification · Model Comparison · Graph Matching · Graph Kernel · Graph Neural Network

1 Introduction

According to recent studies, the machine learning and pattern recognition communities raised concerns about several methodological issues in research. These issues include, for instance, the replicability crisis, which states that previously published results cannot be replicated [1]. Other problems include biased results [2] and models that cannot handle real-world scenarios [3] due to too small datasets or lack of diversity. In summary, poor research practices, such as unclear experimental setups, irreproducible results, and improper model comparisons, hinder consistent evaluations of machine learning and pattern recognition methods and require concerted efforts to prevent their use.

Supported by the Swiss National Science Foundation (SNSF) under Grant Nr. 200021_188496.

A common solution to address improper comparisons of models is to use standardized datasets with rigorous and reproducible experimental designs [4]. In this paper, we provide an exhaustive empirical analysis of three popular graph-based classification methods on 24 node-labeled graph datasets from the TUDataset repository [5]. The analysis includes *graph edit distance* [6], which computes the minimum amount of transformation required to transform one graph into another, a *graph kernel* [7], which computes implicit graph embeddings based on certain graph properties, and finally a *graph neural network* [8], which computes explicit graph embeddings based on the message passing mechanism. These methods are well known and they embody standard classifiers (like *k-NN*, *SVM*, or *Neural Networks*), which are in turn widely used in academic and applied research.

It is, however, not the primary goal of this paper to compare these three graph classifiers. Rather, the goal is to find out in which applications and with which procedures a graph-based approach is actually beneficial at all. To find this out, we test the three above-mentioned classification paradigms in the following three configurations, each representing the same application and setup. **(I)** Graph-based methods using the original graphs (node labeling and graph structure). **(II)** Graph-based methods using the graphs without their labels (only graph structure). **(III)** Vector-based methods which only have access to a label aggregated over all nodes (only node labeling). In particular, the experiments aim at finding out which information is most important in the graph structure (i.e., structure, labeling, or both). Moreover, especially the third configuration can be seen as a naïve baseline, which should be used as a reference system whenever a novel graph-based method is proposed.

We are aware that this is not the first attempt to figure out when, why, and which graph-based methods are actually more valuable than other methods. In [7], for example, a thorough survey of common graph kernels is given, and in [9] it is shown that the results of state-of-the-art graph kernels are sometimes worse than those of some trivial graph kernel methods. In addition, a comprehensive comparison of six graph neural network architectures is presented in [4]. Our analysis differs from previous work in two seminal ways. First, to the best of our knowledge, we are the first to contrast each of the three popular graph-based classifiers with two systems that operate similarly (or identically) but only have access to either the structure or the labels of the graphs. Second, we complete our paper by defining an explicit list of graph datasets that can be used to report the results of (approximate) graph-based methods (or, put negatively, we point out which graph datasets should not be used for such research purposes).

2   Research Context

In Sect. 2.1, we provide a brief overview of the graph classifiers employed in this paper. Additionally, in Sect. 2.2, we discuss the methodical structure of our experiment using the three configurations.

2.1   Graph Classification

The graph classification problem can be addressed through various methods. In this paper, we use three methods based on graph edit distance, graph kernels, and graph neural networks.

Graph Edit Distance [6] is a widely used measure of dissimilarity applicable to any kind of graphs. Graph edit distance basically seeks a sequence of edit operations required to transform a source graph into a target graph. Each of the operations is associated with a cost, and one aims at finding the sequence that minimizes the overall cost. The computation of graph edit distance is an NP-hard problem [10], which means that finding the exact solution for large graphs can be computationally expensive. Therefore, various approximation algorithms to efficiently calculate a suboptimal solution have been proposed. In this paper, we use the popular algorithm *Bipartite-Graph Edit Distance* [11] as a graph edit distance approximation (termed *ged* in this paper). Traditionally, classification based on graph edit distance is limited to the *k-Nearest Neighbor* classifier (*k*-NN).

Graph Kernels [7] represent measures that can be used to quantify the similarity of two graphs. In the present paper, we use the Weisfeiler-Lehman graph kernel [12] (termed *wl* from now on). This popular graph kernel works by assigning each node a label based on the labels of its immediate neighborhood (which is the set of the node's adjacent nodes). This labeling process is iteratively repeated, each time refining the labels based on the updated labels of the neighboring nodes. The advantage of this graph kernel (as well as others) is that it can capture the local and global structural properties of graphs. The actual power of graph kernels is, however, that they implicitly map the graphs into a high-dimensional vector space where the similarity between the graphs can be calculated as the inner product between their corresponding vectors. The resulting implicit embedding can then be fed into a *Support Vector Machine* (SVM) – or any other kernel machine – to perform the final classification.

Graph Neural Networks [8] are designed to learn vectorial representations of nodes, edges, or complete graphs. Graph neural networks gained popularity in recent years due to their ability to handle complex, structured data that is difficult to model using traditional classification methods. The basic idea of graph neural networks is to apply a series of graph convolutional operations that aggregate information from neighboring nodes and edges to update the features of each node. The updated features are then passed through a non-linear activation function and used to predict the target variable. There are different variations of graph neural networks available. In the present paper, we use the *Deep Graph Convolutional Neural Network* (DGCNN) [13] (simply termed *gnn* from now on). This architecture consists of three consecutive stages. First, graph convolutional layers are used to extract local substructure features of the nodes and establish a consistent node ordering. Second, a SortPooling layer arranges the node features in the established order and standardizes input sizes. Third, traditional convolutional and dense layers are utilized to process the sorted graph representations and generate the final classification.

2.2 Classification Methods Comparison

The overall information of a graph consists of two parts, namely the structure and the labels. By omitting one or the other, or using both pieces of information at the same time, we thus obtain three different configurations. We use the three classification paradigms described in the previous subsection in these three configurations.

In Table 1, the three configurations are summarised. **(I)** The initial configuration (shown in column 1) consists of *Labeled Graphs*, which involves conducting graph classification on the original graphs, including both its structure and node labels. **(II)** The second configuration (shown in column 2) aims to examine the significance of the graph structure itself by excluding the node labels to obtain *Unlabeled Graphs*. **(III)** The final configuration (shown in the third column) is the *Aggregated Labels* setup, in which only the node information is retained. We use global sum pooling on the graphs' nodes to obtain a basic vector representation of the graphs. This feature vector is then fed into three statistical classifiers (a k-NN classifier using the Euclidean distance (L_2), an SVM with a radial basis function (rbf), and a multilayer perceptron (mlp)). The rationale of this procedure is to use statistical classifiers that are conceptually closely related to the three graph-based classifiers defined above.

3   Experimental Setup

3.1   Datasets

The following empirical evaluation is conducted on 24 datasets from the TUDataset graph repository [5] that contains various benchmark datasets for graph classification research. For further details of each dataset, including the number of graphs and classes, as well as the average, minimum, and maximum number of nodes and edges per graph, we refer to [5].

Table 1. Three configurations (L, U, A) for the three classifier paradigms (k-NN, SVM, NN) presented. **(I)** Labeled Graphs: Original graphs with node labels and graph structure, **(II)** Unlabeled Graphs: Graphs without labels (graph structure only), and **(III)** Aggregated Labels: Vector representation of the graphs based on the node labels.

	Graph-Based		Vector-Based
Classifier	**Labeled Graphs**	Unlabeled Graphs	**Aggregated Labels**
k-NN	ged(L)	ged(U)	L_2(A)
SVM	wl(L)	wl(U)	rbf(A)
NN	gnn(L)	gnn(U)	mlp(A)

Our study is limited to a set of node-labeled graph datasets. These graphs represent entities obtained from a broad range of fields to provide a comprehensive evaluation across various applications. This particularly includes chemoinformatics, bioinformatics, and computer vision. Roughly speaking, the chemoinformatics datasets consist of molecular graphs, while the bioinformatics datasets comprise protein interaction networks and biological pathways, and the computer vision datasets include image-based graphs.

3.2 Experimental Setup

In order to ensure the reliability of the empirical results and minimize the influence of random data partitionings, we employ a 10-fold cross-validation strategy that is repeated 10 times in a stratified manner for each dataset and configuration. The performance of the models is then estimated using each partition, where hyperparameters are chosen through an internal model selection process that only uses the training data. Note that the model selection is conducted independently for each training and test split, thus the optimal hyperparameter configurations may differ from one split to another.

A common metric to assess the quality of a classifier is the classification accuracy. It measures the relative proportion of correctly classified instances out of the total number of instances. However, the classification accuracy can be misleading in the case of imbalanced datasets, where the number of instances in one class is much larger than the number of instances in another class (this may cause the classifier to predict the majority class more often and thus report over-optimistic results). Therefore, we report the balanced accuracy[14] by default in this paper since not all the datasets used are class balanced.

Concerning the computation of graph edit distance, we use unit costs for both node and edge insertions/deletions. For the node substitution cost, we use the Euclidean distance between the corresponding node labels. For edge substitution, we use a zero cost (since the edges are unlabeled). Parameter $\alpha \in]0, 1[$ represents the relative importance of node and edge edit operation costs and is varied from 0.1 to 0.9 in increments of 0.1 in our evaluation. The only parameter that needs to be optimized for the k-NN classifier (viz. the number of neighbors k) is optimized in the range $k \in \{3, 5, 7\}$.

In the kernel scenario, we use a 4-Weisfeiler-Lehman kernel which means we perform four refinement iterations. To optimize the SVM parameter C, which balances the trade-off between large margins and minimizing misclassification, we explore values in the range $10^{-2.0, -1.5, ..., 2.0}$. We use the same range for the regularization parameter when optimizing the SVM with radial basis function for the aggregated node labels.

For the training process of the gnn experiments, we use the hyperparameters as proposed in [4]. For the fully connected network, the following parameters are optimized. The depth of the standard fully connected layers $d \in \{1, 3, 5, 10\}$, the number of neurons per layer $n \in \{5, 10, 20, 30\}$, the dropout value $\delta \in \{0, 0.25, 0.5\}$ and the learning rate $l \in \{0.005, 0.01, 0.05, 0.1\}$.

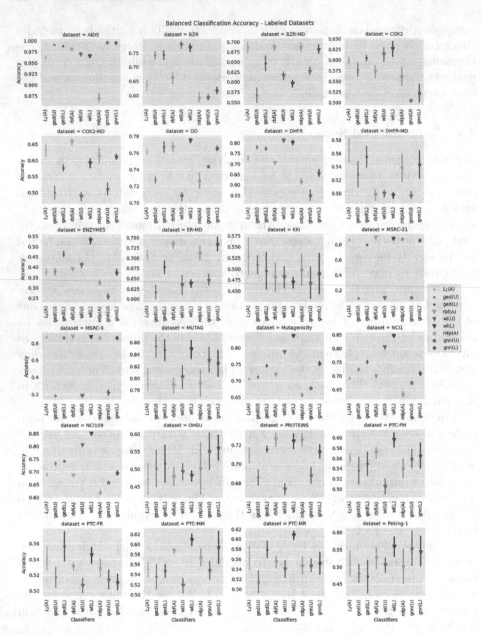

Fig. 1. Balanced classification accuracy of the three types of classifiers (k-NN, SVM, NN) achieved in the three tested configurations (L, U, A) across all datasets.

4   Experimental Evaluation

The experimental evaluation consists of two parts. First, in Sect. 4.1, we compare the classification accuracies achieved using the three classifier paradigms in the three configurations (resulting in nine classifiers according to Table 1). Second, in Sect. 4.2, we analyze on which datasets the graph-based approaches are actually significantly better than the vector-based counterparts.

4.1 Graph Classification

Figure 1 shows the balanced classification accuracy results obtained by the classifiers in the tested settings. For the three graph-based classifiers (ged, wl, gnn) two different configurations are evaluated, viz. labeled graphs (L) and unlabeled graphs (U). The three vector-based classifiers (L_2, rbf, mlp) are tested on the aggregated node labels (A). There are four main trends that can be observed from this plot.

First, we observe that on the datasets Mutagenicity, NCI1, and NCI109 the three different types of classifiers achieve somehow the expected results. That is the statistical classifiers using the aggregated vectors achieve the lowest accuracies, the second-best performance is achieved with graph-based methods using unlabeled graphs, and the best classification results are obtained on the original graphs (that include both structure and node labels).

Second, we notice that on the six datasets BZR-MD, COX2-MD, MSRC-9, MSRC-21, PROTEINS, and DD, the classifiers face difficulties when the node labels are removed from the graphs. That is, on those datasets, the accuracies obtained by the classifiers operating on unlabeled graphs is almost consistently lower than those of the classifiers that operate on the aggregated feature vectors. That indicates that the structure in those datasets is complicated to distinguish from one another, and moreover, that the labels on the nodes play a pivotal role in those applications.

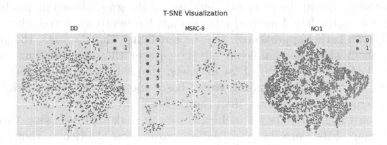

Fig. 2. T-SNE visualization of the vector representation of the graphs for DD, MSRC-9, and NCI1 datasets.

Third, we observe that all classifiers have difficulties performing well on the OHSU, Peking-1, and KKI datasets. The achieved results are only slightly better than random predictions, indicating that these tasks are extremely challenging and currently unsolved by the tested classifiers.

Finally, we observe that on the datasets BZR-MD, COX2, COX2-MD, ER-MD, DHFR-MD, MSRC-9, MSRC-21, DD, PROTEINS, and the four PTCs datasets, the performance of the naïve vector-based approach is comparable to that of the more advanced graph-based techniques. We have two possible explanations for this phenomenon. First, for the MD versions of these datasets, the graphs are heavily modified and fully connected, rendering graph-based classification less effective. Second, by visualizing the aggregated feature vectors with T-SNE (see Fig. 2), we find that the different classes are easily separable, which in turn explains the good results of the vector-based classifiers.

4.2 Dataset Selection

Comparing two classification algorithms is not a trivial task due to the risk of committing type I or type II errors. Type I errors occur when the null hypothesis is wrongly rejected even though it is true, whereas type II errors occur when the null hypothesis is not rejected even though it is false. To address this issue, the authors of [15] conduct an empirical evaluation of multiple statistical tests and conclude that the 10-time repeated 10-fold cross-validation test is the most effective. This test involves all 100 individual systems to estimate the mean and variance of the accuracy with $10°C$ of freedom (making it conceptually simple to use).

We apply this statistical test [15], in order to conduct a comparison between the statistical classifiers that use the aggregated node labeling only and their graph-based counterparts (i.e., the classifiers that use both node labels and graph structure). In particular, this analysis counts how often the graph-based methods outperform their vector-based counterpart.

Regarding the results in Fig. 3, we observe that on six datasets, all of the three graph-based approaches outperform the corresponding vector-based approaches. On four datasets, two of the three graph-based methods outperform the vector-based methods, and in the five cases, only one of the graph-based methods outperforms the vector-based method (this is most frequently observed when comparing the two kernel approaches). On the remaining nine datasets, the graph-based methods show no significant advantage over the corresponding vector-based methods. On the total of 24 datasets, the ged and gnn methods show superiority over the corresponding vector-based methods in 9 cases, while the wl-graph kernel performs better than the rbf-kernel in 14 cases.

Based on the results shown in Fig. 3, we can make the following two recommendations. First, new graph-based approximation algorithms that aim to reduce computation time should only be tested on datasets where the graph-based classification methods outperform the vector-based method in at least two, ideally three, cases. This recommendation is based on the fact that even the fastest approximation method is likely to be slower than the baseline feature vector method presented here. Second, graph-based methods with the goal of improving the classifier rate should be tested primarily on datasets where the graph-based methods evaluated here perform even worse than the corresponding vector-based approaches. This recommendation is based on the fact that on these applications there is still significant potential for improving the structural approaches.

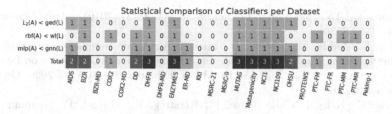

Fig. 3. Comparison of the statistical classifiers that use the aggregated labels (i.e., $L_2(A)$, rbf(A), and mlp(A) and the graph-based approaches that operate on the original graphs with both structure and node labels (i.e., ged(L), wl(L), gnn(L)).

5 Conclusions

The main goal of this paper is to provide a comprehensive analysis of commonly used benchmark datasets for graph-based pattern recognition. More specifically, we investigate on which datasets the structure, the node labels, or both are the most relevant for graph classification. We empirically show the significance of both node labels and structure on various datasets. However, we also reveal that certain common graph classifiers struggle to surpass a basic baseline that uses feature vector representations extracted from the graphs. Based on this analysis, we recommend specific sets of datasets to focus on when evaluating novel (approximation) algorithms related to the three graph-based classifiers used in this paper.

We see three potential ideas for future research activities. First, one could explore whether or not certain graph properties (such as homophily, spectral gap, and others) show a correlation between the classification performances of the different types of classifiers. Second, one could investigate which graph-based classifiers are robust to noise, either in terms of structural changes (e.g., randomly adding or removing edges) or alterations to node information. Third, one could extend our study to unlabeled datasets (our study is limited to graphs with labeled nodes). However, in some preliminary experiments on datasets from social media, we have not yet been able to find unlabeled datasets in which the vector-based approaches outperform the graph-based classifiers. Thus, we conclude that in the case of unlabeled graphs on social media networks, the structural information given by the edges of a graph plays a central role.

References

1. Hutson, M.: Artificial intelligence faces reproducibility crisis. Science **359**(6377), 725–726 (2018). https://doi.org/10.1126/science.359.6377.725
2. Buolamwini, J., Gebru, T.: Gender shades: intersectional accuracy disparities in commercial gender classification. In: Friedler, S.A., Wilson, C. (eds.) Conference on Fairness, Accountability and Transparency, FAT 2018, 23–24 February 2018, New York, USA. Proceedings of Machine Learning Research, vol. 81, pp. 77–91. PMLR (2018)

3. Shmueli, G.: To explain or to predict? Stat. Sci. **25**(3) (2010). https://doi.org/10.1214/10-STS330

4. Errica, F., Podda, M., Bacciu, D., Micheli, A.: A fair comparison of graph neural networks for graph classification. In: 8th International Conference on Learning Representations, ICLR 2020, Addis Ababa, Ethiopia, 26–30 April 2020. OpenReview.net (2020)

5. Morris, C., Kriege, N.M., Bause, F., Kersting, K., Mutzel, P., Neumann, M.: TUDataset: a collection of benchmark datasets for learning with graphs (2020). arXiv: 2007.08663

6. Bunke, H., Allermann, G.: Inexact graph matching for structural pattern recognition. Pattern Recogn. Lett. **1**(4), 245–253 (1983). https://doi.org/10.1016/0167-8655(83)90033-8. May

7. Kriege, N.M., Johansson, F.D., Morris, C.: A survey on graph kernels. Appl. Netw. Sci. **5**(1), 6 (2020). https://doi.org/10.1007/s41109-019-0195-3

8. Yang, Z., et al.: A comprehensive survey of graph-level learning (2023). arXiv: 2301.05860

9. Schulz, T., Welke, P.: On the necessity of graph kernel baselines. In: Graph Embedding and Mining Workshop at ECML PKDD, p. 13 (2019). https://www.semanticscholar.org/paper/On-the-Necessity-of-Graph-Kernel-Baselines-Schulz-Welke/83c7fa53129c983e3124ad0b0d436d982a58c44f

10. Garey, M.R., Johnson, D.S.: Computers and Intractability; A Guide to the Theory of NP-Completeness. W. H. Freeman & Co., USA (1990)

11. Riesen, K., Bunke, H.: Approximate graph edit distance computation by means of bipartite graph matching. Image Vis. Comput. **27**(7), 950–959 (2009). https://doi.org/10.1016/j.imavis.2008.04.004. Jun

12. Shervashidze, N., Schweitzer, P., Leeuwen, E.J.v., Mehlhorn, K., Borgwardt, K.M.J.: Weisfeiler-Lehman graph kernels. Mach. Learn. Res. **12**, 2539–2561 (2011). https://doi.org/10.5555/1953048.2078187

13. Zhang, M., Cui, Z., Neumann, M., Chen, Y.: An end-to-end deep learning architecture for graph classification. In: McIlraith, S.A., Weinberger, K.Q. (eds.) Proceedings of the Thirty-Second AAAI Conference on Artificial Intelligence, (AAAI 2018), the 30th innovative Applications of Artificial Intelligence (IAAI-18), and the 8th AAAI Symposium on Educational Advances in Artificial Intelligence (EAAI 2018), New Orleans, Louisiana, USA, 2–7 February 2018, pp. 4438–4445. AAAI Press (2018)

14. Brodersen, K.H., Ong, C.S., Stephan, K.E., Buhmann, J.M.: The balanced accuracy and its posterior distribution. In: 20th International Conference on Pattern Recognition, ICPR 2010, Istanbul, Turkey, 23–26 August 2010, pp. 3121–3124. IEEE Computer Society (2010). https://doi.org/10.1109/ICPR.2010.764

15. Bouckaert, R.R.: Choosing between two learning algorithms based on calibrated tests. In: Fawcett, T., Mishra, N. (eds.) Machine Learning, Proceedings of the Twentieth International Conference (ICML 2003), 21–24 August 2003, Washington, DC, USA, pp. 51–58. AAAI Press (2003)

A Practical Algorithm for Max-Norm Optimal Binary Labeling of Graphs

Filip Malmberg[1](✉) and Alexandre X. Falcão[2]

[1] Centre for Image Analysis, Department of Information Technology,
Uppsala University, Uppsala, Sweden
`filip.malmberg@it.uu.se`
[2] Institute of Computing, University of Campinas, Campinas, Brazil
`afalcao@ic.unicamp.br`

Abstract. This paper concerns the efficient implementation of a method for optimal binary labeling of graph vertices, originally proposed by Malmberg and Ciesielski (2020). This method finds, in quadratic time with respect to graph size, a labeling that globally minimizes an objective function based on the L_∞-norm. The method enables global optimization for a novel class of optimization problems, with high relevance in application areas such as image processing and computer vision. In the original formulation, the Malmberg-Ciesielski algorithm is unfortunately very computationally expensive, limiting its utility in practical applications. Here, we present a modified version of the algorithm that exploits redundancies in the original method to reduce computation time. While our proposed method has the same theoretical asymptotic time complexity, we demonstrate that is substantially more efficient in practice. Even for small problems, we observe a speedup of 4–5 orders of magnitude. This reduction in computation time makes the Malmberg-Ciesielski method a viable option for many practical applications.

Keywords: Graph labeling · Combinatorial optimization · Lexicographic Max-Ordering

1 Introduction

Many problems in computer science and pattern recognition can be as finding vertex labeling of a graph, such that the labeling optimizes some application-motivated objective function. In their recent work, Malmberg and Ciesielski [9] proposed a quadratic time algorithm for assigning binary labels to the vertices of a graph, such that the resulting labeling is optimal according to an objective function based on the max-norm, or L_∞ norm. Here, we consider the efficient implementation of the algorithm proposed by Malmberg and Ciesielski. We present a version of their algorithm that, while having the same quadratic asymptotic time complexity, is orders of magnitude faster in practice.

A key part of the Malmberg-Ciesielski algorithm is to solve a sequence of *Boolean 2-satisfiability* (2-SAT) problems. Malmberg and Ciesielski observe that

M. Vento et al. (Eds.): GbRPR 2023, LNCS 14121, pp. 35–45, 2023.
https://doi.org/10.1007/978-3-031-42795-4_4

each such 2-SAT problem can be solved in linear time using, e.g., Aspvall's algorithm [1]. They also observe, however, that there is a high degree of similarity between each consecutive 2-SAT problem in the sequence and that solving each 2-SAT problem in isolation thus appears inefficient. Here, we show that this redundancy between subsequent 2-SAT problems can indeed be exploited to formulate a substantially more efficient version of the algorithm.

2   Background and Motivation

We consider the problem of assigning a binary label (0 or 1) to a set of variables identified by indices $1, \ldots, n$. A canonical problem is to find a binary labeling $\ell : [1, n] \to \{0, 1\}$ that minimizes an objective function of the form

$$E_p(\ell) := \sum_i \phi_i^p(\ell(i)) + \sum_{(i,j) \in \mathcal{N}} \phi_{ij}^p(\ell(i), \ell(j)), \tag{1}$$

where $\ell(i) \in \{0, 1\}$ denotes the label of variable i and $\mathcal{N}$ is a set of pairs of variables that are considered *adjacent*.

The functions $\phi_i(\cdot)$ are referred to as *unary* terms. Each unary term depends only on the value of a single binary variable, and they are used to indicate the preference of an individual variable to be assigned each particular label.

The functions $\phi_{ij}(\cdot, \cdot)$ are referred to as *pairwise* terms. Each pairwise term depends on the labels assigned to two variables simultaneously, and thus introduces a dependency between the labels assigned to the variables. Typically, this dependency between variables is used to express that the desired solution should have some degree of smoothness, or regularity.

In applications, rules for assigning these unary and pairwise terms might be hand-crafted, based on the users knowledge about the problem at hand. Alternatively, the preferences might be learned from available annotated data using machine learning techniques [10,11].

As established by Kolmogorov and Zabih [7], the labeling problem described above can be solved to global optimality under the condition that all pairwise terms are submodular, which in the form presented here means that they must satisfy the inequality

$$\phi_{ij}^p(0, 0) + \phi_{ij}^p(1, 1) \leq \phi_{ij}^p(0, 1) + \phi_{ij}^p(1, 0). \tag{2}$$

If the problem contains non-submodular binary terms, finding a globally optimal labeling is known to be NP-hard in the general case [7]. Practitioners looking to solve such optimization problems must therefore first verify that their local cost functional satisfies the appropriate submodularity conditions. If this is not the case, they must resort to approximate optimization methods that may or may not produce satisfactory results for a given problem instance [6]. Recently, however, Malmberg and Ciesielski [9] showed that in the limit case, as p approaches to infinity, the requirement for submodularity disappears! To characterize the labelings that minimize 1 as p goes to infinity, we first observe that as p goes to infinity the objective function E_p itself converges to

$$E_\infty(\ell) := \max\{\max_i \phi_i(\ell(i)), \max_{(i,j)\in\mathcal{N}} \phi_{ij}(\ell(i),\ell(j))\}. \qquad (3)$$

i.e., the objective function becomes the max-norm of the vector containing all unary and pairwise terms. A more refined way of characterizing the solution is the framework of *lexicographic max-ordering* (Lex-MO) [3–5]. The same concept was also studied by Levi and Zorin, who used the term *strict minimizers* [8]. In this framework, two solutions are compared by ordering all elements (in our case, the values of all unary and pairwise terms for a given solution) non-increasingly and then performing their lexicographical comparison. This avoids the potential drawback of the E_∞ objective function, that it does not distinguish between solutions with high or low errors below the maximum error. The Malmberg-Ciesielski algorithm [9] computes, in polynomial time, a labeling that globally minimizes E_∞, even in the presence of non-submodular pairwise terms. Under certain conditions, the same algorithm is also guaranteed to produce a solution that is optimal in the Lex-MO sense. As shown by Ehrgott [4], Lex-MO optimal solutions have the following favorable properties:

- They are *Pareto* optimal, i.e., it is not possible to change the solution to improve one criterion (unary- or pairwise term) without worsening another one.
- They minimize E_∞, i.e., that minimize the largest value of any criterion (unary- or pairwise term).
- All Lex-MO solutions are equivalent in the sense that the corresponding vector of sorted criteria are the same.

3 Preliminaries

In this section, we recall briefly the Malmberg-Ciesielski algorithm, along with some concepts needed for exposition of our proposed efficient implementation of this algorithm in Sect. 4.

3.1 Boolean 2-Satisfiability

We start by recalling the Boolean 2-satisfiability (*2-SAT*) problem. Given a set of Boolean variables $\{x_1,\ldots,x_n\}$, $x_i \in \{0,1\}$ and a set of logical constraints on pairs of these variables, the 2-SAT problem consists of determining whether it is possible to assign values to the variables so that all the constraints are satisfied (and to find such an assignment, if it exists). To formally define the 2-SAT problem, we say that a *literal* is either a Boolean variable x or its negation $\neg x$. A 2-SAT problem can then be defined in terms of a Boolean expression that is a conjunction of *clauses*, where each clause is a disjunction of two literals. Expressions on this form are known as 2-CNF formulas, where CNF stands for *conjunctive normal form.* The 2-SAT problem consists of determining if there exists a truth assignment to the variables involved in a given 2-CNF formula that makes the whole formula true. If such an assignment exists, the 2-SAT problem is said to be *satisfiable*, otherwise it is *unsatisfiable*. As an example, the following expression is a 2-CNF formula involving three variables x_1, x_2, x_3, and two clauses:

$$(x_1 \lor x_2) \land (x_2 \lor \neg x_3) \tag{4}$$

This example formula evaluates to *true* if we, e.g., assign all three variables the value 1 (or *true*). Thus the 2-SAT problem represented by this 2-CNF formula is satisfiable.

For any 2-CNF formula, the 2-SAT problem is solvable in linear time w.r.t to the number of clauses[1] using, e.g., Aspvall's algorithm [1].

We now introduce some further notions related to 2-SAT problems needed for our exposition, using the convention that x_i and $\neg x_i$ denote literals, while v_i denotes a literal whose truth value is unknown and $\bar{v}_i$ is its complementing literal.

Every clause $(v_i \lor v_j)$ in a 2-CNF formula is logically equivalent to an implication from one of its variables to the other:

$$(v_i \lor v_j) \equiv (\bar{v}_i \Rightarrow v_j) \equiv (\bar{v}_j \Rightarrow v_i) . \tag{5}$$

As established by Aspvall et al. [1], this means that every 2-SAT problem F can be associated with an *implication graph* $G_F = (V, E)$, a directed graph with vertices V and edges E constructed as follows:

1. For each variable x_i, we add two vertices named x_i and $\neg x_i$ to G_F. The vertices x_i and $\neg x_i$ are said to be *complementing*.
2. For each clause $(v_i \lor v_j)$ of F, we add edges $(\bar{v}_i, v_j)$ and $(\bar{v}_i, v_j)$ to G_F.

Each vertex in the implication graph can thus be uniquely identified with a literal, and each edge identified with an implication from one literal to another. We will therefore sometimes interchangeably refer to a vertex in the implication graph by its corresponding literal v_i. For a given truth assignment, we say that a vertex in the implication graph *agrees* with the assignment if the corresponding literal evaluates to *true* in the assignment. The implication graph G_F is *skew symmetric* in the sense that if (v_i, v_j) is an edge in G_F, then $(\bar{v}_i, \bar{v}_j)$ is also an edge in G_F. We observe that it follows that for every path $\pi = (v_1, v_2, \ldots, v_k)$ in G_F, the path $\bar{\pi} = (\bar{v}_k, \bar{v}_{k-1}, \ldots, \bar{v}_1)$ is also a path in G_F.

In proving the correctness of our proposed algorithm, we will rely on the following property which is due to Aspvall et al. [1]:

Property 1. *A given truth assignment satisfies a formula F if and only if there is no vertex in G_F for which the corresponding literal agrees with the assignment, with an outgoing edge to a vertex not agreeing with the assignment.*

---

[1] This is in contrast to the general Boolean satisfiability problem, where clauses are allowed to contain more than two literals. Already the 3SAT problem, where each clause can have at most three literals, is NP-hard.

3.2   The Malmberg-Ciesielski Algorithm

For a complete description of the Malmberg-Ciesieleski algorithm, we refer the reader to the original publication ([9], Algorithm 1). We focus here on a key aspect of the algorithm, which is to solve a sequence of 2-SAT problems. In this step, we identify the variables to be labeled with the Boolean variables involved in a 2-SAT problem. A truth assignment T for the Boolean variables naturally translates to a labeling ℓ. For this step of the algorithm, we are given an ordered sequence $\mathcal{C}$ of clauses, ordered by a priority derived from the unary and pairwise terms in Eq. 3. Informally, the algorithm operates as follows:

- Initialize F to be an empty 2-CNF formula, containing no clauses.
- For each clause c in $\mathcal{C}$, in order:
  - If $F \wedge c$ is satisfiable, then set $F \leftarrow F \wedge c$.

At all steps of the above algorithm, the formula F remains satisfiable. At the termination of the algorithm, the formula F defines a unique truth assignment T and therefore also a labeling ℓ. For the specific sequence $\mathcal{C}$ of clauses defined by Malmberg and Ciesieleski, the resulting labeling is guaranteed to globally minimize the objective function in Eq. 3.

In each iteration, we need to determine if $F \wedge c$ is satisfiable, i.e., solve the 2-SAT problem associated with the formula $F \wedge c$. Malmberg and Ciesieleski suggest to use Aspvall's algorithm for this purpose, with an asymptotic time complexity of $\mathcal{O}(|F|) \leq \mathcal{O}(|\mathcal{C}|)$. Let $N = n + |\mathcal{N}|$ denote the total number of unary and pairwise terms in Eq. 3. By its design, the number of clauses in the sequence $\mathcal{C}$ is $\mathcal{O}(N)$, leading to the asymptotic time complexity of $\mathcal{O}(N^2)$ for the Malmberg-Ciesielski algorithm implemented using Aspvall's algorithm.

4   Proposed Algorithm

As observed in the previous section, the Malmberg-Ciesielski algorithm iteratively builds a formula F that remains satisfiable at each step of the algorithm. Our approach for improving the efficiency of the computations is to maintain, at each step of the algorithm, a truth assignment that satisfies the current formula F. When trying to determine whether the next clause c in the sequence C can be appended to F without rendering the formula unsatisfiable, we show that this previous truth assignment can be utilized to reduce the computation time. We represent a truth assignment T to the Boolean variables of a 2-SAT problem as a function $T : [1, n] \rightarrow \{0, 1\}$, so that $T(i)$ is the value assigned to variable x_i. Trivially, if T satisfies c then is also satisfies $F \wedge c$, so we focus on the case where T does not satisfy the next clause c.

We will consider 2-SAT-solving under *assumptions* [2], i.e., given a satisfiable formula, we ask if the same formula still satisfiable if we assume given values for a subset of the variables? Such assumptions will be represented by a set of vertices in the implication graph – since each vertex corresponds to a literal, the set of vertices corresponds to a set of literals that are all assumed to evaluate to *true.* We assume that vertex sets used in this context are internally conflict-free, i.e., they do not contain both a vertex and its complement.

Below we will present an efficient algorithm for solving a 2-SAT problem under a set of assumptions A, given a truth assignment T that satisfies the formula *without* the assumptions. To see how such a procedure helps us in efficiently implementing the Malmberg-Ciesielski algorithm, we observe that by De Morgan's laws a clause $(v_i \lor v_j)$ can be rewritten as $\neg(\bar{v}_i \land \bar{v}_j)$. In this form, it is easier to see that in order to satisfy this clause, the truth assignment T must satisfy exactly one of the expressions $(v_i \land v_j)$, $(v_i \land \bar{v}_j)$, or $(\bar{v}_i \land v_j)$. Each of these expressions represent a set of assumptions, and therefore $F \land (v_i \lor v_j)$ is satisfiable if and only if F is satisfiable under one of the following sets of assumptions A: $\{v_i, v_j\}$, $\{v_i, \bar{v}_j\}$, or $\{\bar{v}_i, v_j\}$. We note also that in the special case that $i = j$, the above argument can be simplified further. In this case, the formula reduces to $F \land (v_i)$ which is equivalent to solving F under the assumption $A = \{v_i\}$.

The procedure listed in Algorithm 1 utilizes this result to perform the inner loop of the Malmberg-Ciesieleski algorithm: It determines whether a given clause can be added to a satisfiable formula without making it unsatisfiable. If so, it updates an implication graph representing the formula to include the new clause. Algorithm 1 utilizes a procedure *SolveWithAssumptions*, which we will now describe.

Let F be a formula with corresponding implication graph $G_F = (V, E)$, let T be a truth assignment for the variables associated with F, and let A be a set of assumptions. We define $R_{A,T} \subseteq V$ as the set of vertices that are reachable in G_F from any vertex in A without traversing an edge that is outgoing from a vertex that agrees with T. The main theoretical result that enables our proposed algorithm is summarized in the following theorem:

Theorem 1. *Assume that F is satisfiable. Let T be a truth assignment that satisfies F, and let A be a set of assumptions. Then F is satisfiable under the assumptions A if and only if the subgraph $R_{A,T}$ does not contain a pair of complementing vertices.*

Proof. For the first part of the proof, assume that $R_{A,T}$ does contain a pair of complementing vertices v_i and $\bar{v}_i$. Then the assumptions A directly imply that both v_i and $\bar{v}_i$ are simultaneously satisfied, which is clearly a contradiction, and so F is not satisfiable under the assumptions A.

For the second part of the proof, assume that $R_{A,T}$ does not contain any pair of complementing vertices. We may then construct a well-defined truth assignment T' from the given truth assignment T by setting, for every vertex in $R_{A,T}$, the corresponding variable to the corresponding truth value. For any vertex $v_i \notin R_{A,T}$, we have $T(i) = T'(i)$. Furthermore, the truth assignment T' agrees with all assumptions in A.

Next assume, with the intent of constructing a proof by contradiction, that the truth assignment T' constructed above does not satisfy F. Then by Property 1 there exists at least one vertex v_i agreeing with T' that has an outgoing edge to a vertex v_j not agreeing with T'. We now consider all four possibilities

---

Algorithm 1: CheckSolvable(G,C,T)

Input: An implication graph G representing a 2-SAT problem. A clause
$c = (v_i) \vee (v_j)$. A truth assignment T that satisfies the formula F
encoded by G.

Result: A truth value indicating if $F \wedge c$ is satisfiable. If it is, then T is a
truth assignment satisfying $F \wedge c$ and G encodes $F \wedge c$. Otherwise, T
and G are unmodified.

1   Set *satisfiable* $\leftarrow$ *false*
2   **if** T *satisfies* c **then**
3 | Set *satisfiable* $\leftarrow$ *true*

4   **else**
5 | **if** $v_i = v_j$ **then**
6 | **if** *SolveWithAssumptions(G,$\{v_i\}$,T)* **then**
7 | Set *satisfiable* $\leftarrow$ *true*

8 | **else**
        /* $v_i \neq v_j$ */
9 | **if** *SolveWithAssumptions(G,$\{v_i,v_j\}$,T)* **then**
10 | Set *satisfiable* $\leftarrow$ *true*
11 | **else if** *SolveWithAssumptions(G,$\{\bar{v}_i,v_j\}$,T)* **then**
12 | Set *satisfiable* $\leftarrow$ *true*
13 | **else if** *SolveWithAssumptions(G,$\{v_i,\bar{v}_j\}$,T)* **then**
14 | Set *satisfiable* $\leftarrow$ *true*

15   **if** *satisfiable* **then**
16 | Add edges $(\bar{v}_i, v_j)$ and $(\bar{v}_j, v_i)$ to G
17   Return *satisfiable*

---

for the truth assignment T with respect to the variables corresponding to v_i and
v_j :

1. Assume that both v_i and v_j agree with T. Then since v_j does not agree with
   T' we must have $\bar{v}_j \in R_{A,T}$, i.e., there exists a path π from A to $\bar{v}_j$ that does
   not traverse an edge outgoing from a vertex that agrees with T. By the skew
   symmetry of the implication graph, there is an outgoing edge from $\bar{v}_j$ to $\bar{v}_i$,
   and we may thus append this edge to the path π to see that $\bar{v}_i$ is also in $R_{A,T}$,
   contradicting that v_i agrees with T'. Thus, the assumption that both v_i and
   v_j agree with T leads to a contradiction.
2. Assume that v_i agrees with T but v_j does not. Since v_i has an outgoing edge
   to v_j, this contradicts that T satisfies F, and so the assumption that v_i agrees
   with T but v_j does not agree with T leads to a contradiction.
3. Assume that v_j agrees with T but v_i does not. Then v_i and $\bar{v}_j$ are both
   in $R_{A,T}$. There is an outgoing edge from v_i to v_j, and v_i disagrees with T,
   and thus v_j is also in $R_{A,T}$, contradicting the assumption that $R_{A,T}$ does

---

Algorithm 2: SolveWithAssumptions(G,A,T)

Input: An implication graph G representing a 2-SAT problem. A set of
assumptions A, without internal conflicts. A truth assignment T that
satisfies the formula F encoded by G.

Result: A truth value indicating the existence of a truth assignment T' that
satisfies the formula F encoded by G while simultaneously satisfying
the assumptions A. If the algorithm returns *true*, then T is a truth
assignment satisfying this criterion. Otherwise, T is unmodified.

Auxiliary: A FIFO (or LIFO) queue Q of vertices; A set of vertices C.

1 Set $C \leftarrow \emptyset$
2 **foreach** $v \in A$ **do**
3 | Insert v in Q
4 |___ Insert v in C

5 **while** Q *is not empty* **do**
6 | Pop a vertex v from Q
7 | **if** v *disagrees with* T **then**
8 | | **foreach** *vertex w such v has an outgoing edge to w* **do**
9 | | | **if** $\bar{w} \in C$ **then**
10 | | | |___ Return *false* and exit
11 | | | **else if** $w \notin C$ **then**
12 | | | | Insert w in Q
13 | | | |___ Insert w in C

14 **foreach** *vertex* $v \in C$ **do**
15 |___ Set value of T for the variable corresponding to v so that it agrees with v.

16 Return *true*

---

not contain both a vertex and its complement. Thus, the assumption that v_j
agrees with T but v_i does not agree with T leads to a contradiction.

4. Assume that neither v_i nor v_j agree with T. Then since v_i agrees with T' we
must have $v_i \in R_{A,T}$, i.e., there exists a path π from A to v_i that does not
traverse an edge outgoing from a vertex that agrees with T. But since there is
an outgoing edge from v_i to v_j and v_i does not agree with T, we may append
π with this edge to see that v_j must also be in $R_{A,T}$, contradicting that v_j
disagrees with T'. Thus, the assumption that neither v_i nor v_j agree with T
leads to a contradiction

The four cases above cover all possible configurations for the truth values of
the variables corresponding to v_i and v_j in the truth assignment T, and each
case leads to a contradiction. We conclude that the assumption that T' does not
satisfy F leads to a contradiction, and thus T' must satisfy F. This completes
the proof.   □

Based on the theorem presented above, we can solve a 2-SAT problem under
given assumptions if we can find the set $R_{A,T}$. We observe that for a given

set of assumptions, the set $R_{A,T}$ can easily be found in $\mathcal{O}(V + E)$ time using, e.g., breadth-first search. If we, during this breadth-first search, encounter a vertex whose complement is already confirmed to be in $R_{A,T}$, we may terminate the search and return *false*. Pseudocode for this approach is presented in Algorithm 2. With an upper bound of $\mathcal{O}(V + E)$ for solving each 2-SAT problem, the proposed approach has the same quadratic asymptotic time complexity as the approach using Aspvall's algorithm. In practice, however, we will see that the set $R_{A,T}$ is a very small subset of the implication graph, making this approach much faster than running Aspvall's algorithm for every iteration of the Malmberg-Ciesielski algorithm.

In terms of memory complexity, both our proposed approach and Aspvall's algorithm operate on the implication graph. The number of vertices in this graph is two times the number of unary terms, and the number of edges is up to four times the number of pairwise terms. Thus, the memory complexity of the proposed approach (and the approach using Aspvall's algorithm) is linear in the number of binary and pairwise terms.

5   Evaluation

To evaluate the performance of our proposed version of the Malmberg-Ciesielski to the original formulation using Aspvall's algorithm, perform an empirical study emulating a typical optimization scenario in image processing and computer vision. We perform binary labeling of the pixels of a 2D image of size $W \times H$. The neighborhood relation $\mathcal{N}$ is defined by the standard 4-connectivity used in image processing. Values for the unary and pairwise terms are drawn randomly from a uniform distribution. We then compare the computation time of the two implementations, for image sizes varying from 8×8 to 64×64. We only measure the time required for solving the sequence of 2-SAT problems, as this is the only aspect that differs between the implementations. The results are shown in Fig. 1. As the figure shows, the computation time for the implementation based on Aspvall's algorithm increases dramatically with increasing problem size. For an image of size 64×64, the implementation based on Aspvall's algorithm runs in $62\,\text{s}$, while the proposed implementation only requires $0.004\,\text{s}$ for the same computation – a speedup of more than four orders of magnitude.

To further study the computation time of the proposed implementation with respect to problem size, we perform a separate experiment on images with sizes varying from 128×128 to 4096×4096, for which the implementation using Aspvall's algorithm becomes prohibitively slow. The results are shown in Fig. 2. As can be seen from the figure the empirical relation between problem size and computation time appears closer to a linear function across this range, rather than quadratic relation suggested by the worst-case asymptotic time complexity.

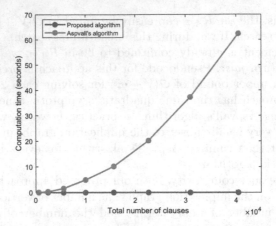

Fig. 1. Comparison of computation time between the proposed implementation of the Malmberg-Ciesielski method, and the original formulation using Aspvall's algorithm, with respect to the total number of clauses in the 2-SAT sequence.

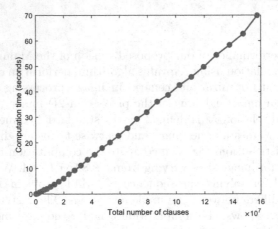

Fig. 2. Computation time of the proposed implementation in relation to problem size.

6   Conclusions

We have proposed a modified, efficient implementation of the Malmberg-Ciesielski method for optimal binary labeling of graphs. While our proposed implementation has the same asymptotic run-time complexity as the original algorithm, we demonstrate that it is orders of magnitude faster in practice. This reduction in computation time makes the Malmberg-Ciesielski method a viable option for many practical applications.

Acknowledgment. This work was supported by a SPRINT grant (2019/08759-2) from the São Paulo Research Foundation (FAPESP) and Uppsala University.

References

1. Aspvall, B., Plass, M.F., Tarjan, R.E.: A linear-time algorithm for testing the truth of certain quantified boolean formulas. Inf. Process. Lett. **8**(3), 121–123 (1979)
2. Eén, N., Sörensson, N.: An extensible SAT-solver. In: Giunchiglia, E., Tacchella, A. (eds.) SAT 2003. LNCS, vol. 2919, pp. 502–518. Springer, Heidelberg (2004). https://doi.org/10.1007/978-3-540-24605-3_37
3. Ehrgott, M.: Lexicographic max-ordering-a solution concept for multicriteria combinatorial optimization (1995)
4. Ehrgott, M.: A characterization of lexicographic max-ordering solutions (1999). http://nbn-resolving.de/urn:nbn:de:hbz:386-kluedo-4531
5. Ehrgott, M.: Multicriteria optimization, vol. 491. Springer Science & Business Media (2005)
6. Kolmogorov, V., Rother, C.: Minimizing nonsubmodular functions with graph cuts-a review. IEEE Trans. Pattern Anal. Mach. Intell. **29**(7) (2007)
7. Kolmogorov, V., Zabih, R.: What energy functions can be minimized via graph cuts? IEEE Trans. Pattern Anal. Mach. Intell. **26**(2), 147–159 (2004)
8. Levi, Z., Zorin, D.: Strict minimizers for geometric optimization. ACM Trans. Graph. (TOG) **33**(6), 185 (2014)
9. Malmberg, F., Ciesielski, K.C.: Two polynomial time graph labeling algorithms optimizing max-norm-based objective functions. J. Math. Imaging Vision **62**(5), 737–750 (2020)
10. Wolf, S., et al.: The mutex watershed and its objective: efficient, parameter-free graph partitioning. IEEE Trans. Pattern Anal. Mach. Intell. **43**(10), 3724–3738 (2020)
11. Wolf, S., Schott, L., Kothe, U., Hamprecht, F.: Learned watershed: End-to-end learning of seeded segmentation. In: Proceedings of the IEEE International Conference on Computer Vision, pp. 2011–2019 (2017)

An Efficient Entropy-Based Graph Kernel

Aymen Ourdjini[1], Abd Errahmane Kiouche[2], and Hamida Seba[2]($\boxtimes$)

[1] École nationale Supérieure d'Informatique (ESI), Oued Smar, Alger, Algérie
ga_ourdjini@esi.dz
[2] Université de Lyon, Université Lyon 1, LIRIS UMR 5205, 69622 Lyon, France
{abd-errahmane.kiouche,hamida.seba}@univ-lyon1.fr

Abstract. Graph kernels are methods used in machine learning algorithms for handling graph-structured data. They are widely used for graph classification in various domains and are particularly valued for their accuracy. However, most existing graph kernels are not fast enough. To address this issue, we propose a new graph kernel based on the concept of entropy. Our method has the advantage of handling labeled and attributed graphs while significantly reducing computation time when compared to other graph kernels. We evaluated our method on several datasets and compared it with various state-of-the-art techniques. The results show a clear improvement in the performance of the initial method. Furthermore, our findings rank among the best in terms of classification accuracy and computation speed compared to other graph kernels.

Keywords: Graph Kernels · Graph Entropy · Graph similarity

1 Introduction

Graph kernels have emerged over the past two decades as promising approaches for machine learning on graph-structured data [15]. Numerous graph kernels have been proposed to address the problem of evaluating graph similarity, enabling predictions in classification and regression contexts [3]. All graph kernel methods adhere to a consistent two-step process [14]. The initial step entails proposing a kernel function to compute the similarity between graphs. This is achieved by mapping the graphs into an alternate vector space, known as the feature space, where similarities are then calculated. Subsequently, the derived function is employed to generate the graph kernel matrix, which contains the similarity values between every pair of graphs. The second stage involves applying a machine learning algorithm to ascertain the optimal separation geometry within the feature space generated during the first step. The primary challenge of graph kernels is to be both expressive and efficient in accurately measuring the similarity between graphs. These kernels implicitly project graphs into a feature space, denoted as H. Following this projection, the scalar product within the feature space is calculated to measure the similarity.

© The Author(s), under exclusive license to Springer Nature Switzerland AG 2023
M. Vento et al. (Eds.): GbRPR 2023, LNCS 14121, pp. 46–56, 2023.
https://doi.org/10.1007/978-3-031-42795-4_5

Numerous graph kernel computation methods have been proposed in the literature, featuring similarity calculation principle based on various concepts such as random walks [9], shortest paths [4], substructure enumeration [19], etc. In this work, we focus on a relatively recent approach of new graph kernels based on entropy [21,22]. In information theory, entropy measures the amount of information contained in an information source. In a graph, it quantifies the information stored within the graph, essentially capturing the complexity and organization level of the graph's structural features [1,2]. In other words, it measures the amount of structural information (or structural complexity) generated by a graph; the larger its value, the more complex the graph and the more diverse its structure. Our analysis of entropy-based graph kernels, especially the Renyi entropy-based graph kernel, introduced in [21], revealed several weaknesses such as redundant computations that slows the algorithm and a lack of accuracy mainly related to the fact that this kernel does not take into account node attributes.

In this work, we address these weaknesses through a new graph kernel based on the concept of Von Neumann entropy. Von Neumann entropy [12] allows us to consider both structural information and attributes, thereby achieving better accuracy. We also propose a new strategy for calculating similarity scores between two graphs, which is significantly faster. Our proposed kernel ranks among the fastest in terms of execution time. The remainder of the paper is organized as follows: In Sect. 2 we analyze related work and motivate our contribution. In Sect. 3, we describe the new proposed kernel. Section 4 presents the experiments that we undertook to assess the efficiency of the proposed method.

2   Related Work and Motivation

Numerous graph kernels have been proposed in the literature. A detailed study of existing approaches can be found in [11]. These techniques aim to measure the similarity between graphs by assessing the relationships between their constituent elements. We will outline the primary approaches and their basic principles, focusing on the granularity of the graph elements considered in each case.
Node-Centric Approaches: These methods assess the similarity between two graphs by examining the similarities between their nodes. Each node is assigned a simple descriptor that encapsulates information from its direct or extended neighborhood, which could be in the form of labels [20], vectors [13], or embeddings generated using neural networks. The resulting node representations are compared using a base kernel function. Node-centric approaches are computationally efficient and relatively simple, making them particularly suited for large-scale graph data. Notable examples include Neighborhood Hash Kernel [8], Weisfeiler-Lehman graph kernels [17,18], Renyi entropy-based kernel [21], and optimal assignment kernels between nodes, such as WL-OA [16].
Subgraph-Oriented Approaches: These techniques represent graphs as sets of subgraphs and measure similarity by comparing the substructures present in each

graph. By decomposing graphs into their constituent subgraphs, these methods can capture more complex relationships between graph structures. Prominent examples include Graphlet Sampling Kernel [19] and Neighborhood Subgraph Pairwise Distance Kernel (NSPD) [5]. While subgraph-oriented approaches offer richer graph representations, they may be computationally intensive, particularly when dealing with large graphs or a diverse range of subgraph types.

Path and Walk-Based Approaches: These kernels compare sequences of node or edge attributes encountered during graph traversals. The similarity between two graphs is quantified by assessing the similarity of their traversal sequences. Path and walk-based approaches are divided into two main groups. The first group focuses on the comparison of shortest paths, including Shortest Path Kernel [4] and Graph Hopper Kernel [7]. The second group is based on random walk comparisons, such as Random Walk Graph Kernel [9]. These methods are well-suited for capturing topological and structural patterns in graphs.

Numerous graph kernel approaches have been proposed, each with its own strengths and weaknesses. In the context of our study, we are particularly interested in recent advances in entropy-based graph kernels [21, 22]. These methods leverage information theory to quantify the complexity and organization of a graph's structural features, thus providing a powerful means of capturing graph similarity.

The graph kernel proposed by Xu et al. [21] is the most recent approach based on the concept of graph entropy. In this method, the authors use Renyi entropy [6] to compute similarities between graphs. The Renyi entropy of a graph $G = (V, E)$ is given by Eq. 1, where the probability $P_G(v)$ represents the ratio of the degree of node v to the sum of degrees of all nodes in the graph.

$$E_{Renyi}(G = (V, E)) = -log(\sum_{v \in V} P_G(v)^2) \qquad (1)$$

The authors then opted for a multi-layer vector representation of entropy for each node in the graph. The fundamental idea is to represent each node by a vector containing the entropies of subgraphs (layers) at different depths surrounding the node. Figure 1 illustrates an example of the representation used for each node in the graph. It is evident that if two nodes in different graphs have similar vector representations, the structures (subgraphs) surrounding them are similar since they have the same structural complexity score (Renyi entropy). Finally, the similarity score (kernel) is calculated by solving an alignment problem between the vector representations of the nodes in the compared graphs. However, this approach exhibits the following weaknesses:

1. The multi-layer representation of each node generates information redundancy that may bias the similarity score between two graphs. Indeed, the subgraph of order i surrounding a given node is entirely included in the higher order subgraphs. This causes the entropy scores of different layers to be implicitly correlated to some extent. Figure 1 illustrates a visualization of such a situation, where an order i subgraph always constitutes part of the next order, i;e., $i + 1$, subgraph. In this example, the entropy scores of the three layers are very close, resulting in information redundancy.

2. Renyi entropy [6] (see Eq. 1) is simple and exploits only the distribution of node degrees in a subgraph. This could negatively impact the accuracy of the similarity measure. Indeed, the utilized entropy depends solely on node degrees without considering neighborhood relationships, making it a coarse characterization of graphs. Moreover, there are numerous graphs with exactly the same degrees but different structures.

3. The entropy formula used does not take into account node labels (types) or attributes. However, these two pieces of information are crucial in many application domains, particularly in biology and chemistry.

4. The assignment step, which calculates the kernel between two graphs using the node vectors, is time-consuming. Furthermore, this assignment does not take into account node labels (types) or attributes.

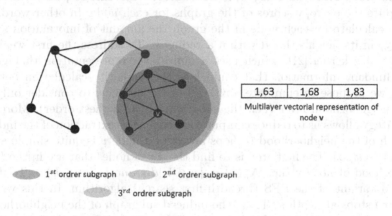

| 1,63 | 1,68 | 1,83 |
Multilayer vectorial representation of node v

1st ordrer subgraph 2nd ordrer subgraph

3rd ordrer subgraph

Fig. 1. Mutli-Layer representation [21]

In this work, we propose a new graph kernel based on the concept of Von Neumann entropy [12]. Our approach extends and improves the method proposed by Xu et al. [21], addressing all the previously mentioned weaknesses. Furthermore, unlike the method of Xu et al., our kernel can be applied to labeled or attributed graphs.

3   Von Neumann Entropy Based Graph Kernel

Algorithm 1 shows the main steps of our method. Our method takes as input two labeled or attributed graphs, $G_1 = (V_1, E_1, \mathcal{L}_1)$ and $G_2 = (V_2, E_2, \mathcal{L}_2)$. The first step consists of extracting the induced subgraphs of the neighborhood of each node in both graphs. Next, we compute the entropy score of each node in both graphs by applying Von Neumann entropy [12] that allows taking into account the types (labels) or attributes of the nodes. The final step involves calculating the kernel between the two graphs based on the entropy scores of the nodes in the two graphs to be compared. In the following, we will detail each of these steps.

Algorithm 1: Von-Neumann Entropy based Graph Kernel

   Input: Graphs G_1 and G_2
   Output: Von-Neumann entropy graph kernel $K(G_1, G_2)$
 1 Extract induced neighborhood subgraph for each node in G_1 and G_2
 2 Compute entropy scores for each subgraph
 3 Compute the kernel $K(G_1, G_2)$ using the entropy scores
 4 **return** $K(G_1, G_2)$

Neighborhood Induced Subgraph Extraction: The first step of our kernel is to compute the entropy scores of the graphs for each node. In other words, we want to calculate for each node in the graph the amount of information stored (entropy) in its neighborhood within a radius k. To address the first weakness of Xu et al.'s kernel [21], which uses a multi-layer representation that generates redundant information that could bias the similarity calculation between graphs, we propose to eliminate any kind of redundancy, to consider only the largest neighborhood layer (i.e., the subgraph of the highest order). Moreover, this strategy allows us to reduce computation time. The extraction of the induced subgraph of the neighborhood (k-hops away) of a node v is quite simple and is done in two steps. The first step is to find the set of nodes that are in the k-hop neighborhood of node v (i.e., $\mathcal{N}_k(v) = \{u \in V | distance(u,v) \leq k\}$). To do so, we use a variant of the BFS (breadth-first search) algorithm. In this variant, the search stops at depth k. Then, the induced subgraph of the neighborhood is constructed from the edges of the graph and the set of nodes in the k-hop neighborhood $N_k(v)$. The radius-k neighborhood induced subgraph around vertex v is $\zeta_v^k(\mathcal{V}_v^k, \varepsilon_v^k)$, with:

$$\mathcal{V}_v^k = \{u \in \mathcal{N}_k(v)\} \tag{2}$$

$$\varepsilon_v^k = \{(u,v) \subset \mathcal{N}_k(v) | (u,v) \in E\} \tag{3}$$

The second step involves computing the amount of information contained in the neighborhood of each node. This amount is used to compute the entropies of the induced subgraphs constructed in the first step. To address the second and third weaknesses of the kernel by Xu et al. [21], we will use an adapted variant of Von Neumann entropy [12], which takes into account node types (labels) and attributes in the graph. This is calculated with respect to a node $v \in V$, called the "root node." We adapt the Von Neumann formula [12] so that the entropy value of a graph captures the similarities of labels (types) or attributes of the node. To do this, we consider only the neighbours in the calculation of the entropy of a neighborhood subgraph $\zeta_v^k(\mathcal{V}_v^k, \varepsilon_v^k)$ of a node v that have the same type (label) or attribute as the root node v. Suppose that the set of labels (types) or attributes of the nodes is $\mathcal{L}$, and $\ell : \mathcal{V}_v^k \to \mathcal{L}$ is a function that assigns labels (or attributes) to the graph vertices. The set $\mathcal{V}_{\ell(v)}^k$ represents the nodes of G that have the same label (or attribute) as the root v, i.e., $V_{\ell(v)} = u \in \mathcal{V}_{\ell(v)}^k : \ell(u) = \ell(v)$. $d = |\mathcal{V}_{\ell(v)}^k|$ is the cardinality of $\mathcal{V}_{\ell(v)}^k$. The entropy that we define for the subgraph ζ_v^k with respect to the root v is then equal to:

$$H_\ell(\zeta_v^k, v) = 1 - \frac{1}{d} - \sum_{u \in \mathcal{V}_{\ell(v)}^k} \frac{1}{d^2 \times deg(v) \times deg(u)} \tag{4}$$

At the end of this step, we represent each graph $G = (V, E, \mathcal{L})$ by a vector $D_k(G)$ containing the entropy scores of each node $v \in V$ along with its label or attribute.

$$D_k(G) = [(H_\ell(\zeta_{v_1}^k, v_1), \ell(v_1)), (H_\ell(\zeta_{v_2}^k, v_2), \ell(v_2)), \cdots, (H_\ell(\zeta_{v_n}^k, v_n), \ell(v_n))] \tag{5}$$

Kernel Calculation: The final step involves calculating the kernel (similarity) between the graphs. Let G_1 and G_2 be two graphs, and D_1, D_2 their vector representations obtained from the previous step. Our entropy kernel between G_1 and G_2 is defined as follows:

$$K_{LE}(G_1, G_2) = K(D_1, D_2) = \sum_{e_1 \in D_2} \sum_{e_2 \in D_2} k_e(e_1, e_2) \tag{6}$$

where $k_e(e_i, e_j)$ is a positive semi-definite kernel on the entropies of the nodes. This kernel is designed to compare both the entropy values and the labels (attributes) of their root nodes. Let $e_1 = (H_{\ell 1}, \ell_1)$ and $e_2 = (H_{\ell 2}, \ell_2)$. H_ℓ represents the entropy value and ℓ denotes the label (or attribute). In the case of labeled graphs, the kernel k_e is defined as follows:

$$k_e(e_1, e_2) = \begin{cases} 1, & \text{if } H_{\ell 1} = H_{\ell 2} \text{ and } \ell_1 = \ell_2 \\ 0, & \text{otherwise} \end{cases} \tag{7}$$

Finally, we apply the RBF (Radial Basis Function) kernel to calculate the final similarity kernel. This combination with RBF is chosen based on several research works [11, 15] which have found that applying RBF leads to a significant improvement in the accuracy of several graph kernels. The final value of our kernel called (Von-Neumann Entropy Graph Kernel) **VEGK** between two graphs is given by the following formula, where σ is a positive real number, its value is generally determined by calibration [15].

$$K_{VEGK}(G_1, G2) = exp\left(-\frac{K_{LE}(G_1, G_1) - 2K_{LE}(G_1, G_2) + K_{LE}(G_2, G_2)}{\sigma^2}\right) \tag{8}$$

4   Evaluation

We evaluated the performance of our graph kernel on 10 publicly accessible datasets [10]. These datasets come from different fields, including cheminformatics, bioinformatics, and social networks. All graphs are undirected. Moreover, the graphs contained in the cheminformatics and bioinformatics datasets have labeled nodes, attributed nodes, or both. Table 1 presents the main characteristics of the datasets. The "Class Imbalance" column indicates the ratio between the size of the smallest class in the dataset and the size of its largest class. (Num.) denotes the number of labels contained in the set, while (Dim.) indicates the dimension of the attributes. The symbol $(-)$ indicates that the graphs in the set do not contain node labels or attributes.

Table 1. Summary of the 10 datasets used in our experiments

Dataset	Statistics				Nodes labeling	
	# Graphs	Classes	Avg V	Avg E	Labels (Num.)	Attributes (Dim.)
MUTAG	188	2	17.93	19.79	+ (7)	–
AIDS	2000	2	15.59	16.2	+ (38)	+ (4)
ENZYMES	600	6	32.46	62.14	+ (3)	+ (18)
MSRC_21C	209	17	40.28	96.6	+ (21)	–
PROTEINS	1113	2	39.05	72.82	+ (3)	+ (1)
PTC_MR	344	2	14.29	14.69	+ (18)	–
IMDB-BINARY	1000	2	19.77	96.53	–	–
IMDB-MULTI	1500	3	13.0	65.94	–	–
SYNTHETICnew	300	2	100.0	196.25	-	+ (1)
Synthie	400	4	91.6	172.93	–	+ (15)

"–" means no node labels

We rely on two main criteria: The accuracy of the kernel (Accuracy) and the computation time (measured in seconds). Our approach is compared to the Renyi entropy kernel (Second-order Rényi Entropy Graph Kernel) **SREGK** proposed in [21] as well as several other existing graph kernels: (1) Shortest path kernel (**SP**) [4], (2) Graph Hopper Kernel (**GH**) [7], (3) Random Walk Kernel (**RW**) [9], (4) Graphlet Sampling kernel (**GS**) [19], (5) Neighborhood Hash Kernel (**NH**) [8], (6) Weisfeiler-Lehman Optimal Assignment (**WL-OA**) [18], and (7) Neighborhood Subgraph Pairwise Distance (**NSPD**) [5].

In our experiments, we use 10-fold cross-validation by applying the C-Support Vector Machine (C-SVM) classification to calculate the classification accuracy. We used nine out of ten samples for training and one sample for testing. We calibrated the parameters of each of the methods mentioned above on each dataset.

Results on Labeled/Attributed Graphs: Table 2 shows the computation times of the kernel matrices on the 6 datasets containing labeled graphs. Table 3 illustrates the classification accuracy scores of all considered kernels on the 6 labeled/attributed datasets. We observe that our kernel is the fastest on 5 datasets. On the MSRC-21 dataset, our kernel ranks 2nd just after the **NH** kernel, with the difference between the two not being very significant. In terms of classification accuracy, our kernel did not achieve the best performance, but it proved competitive compared to other kernels. We can observe that on all datasets except AIDS, our approach outperforms the **SREGK** approach. This demonstrates the usefulness and effectiveness of the improvements we have proposed.

Table 2. Average running time of kernel computation over the 6 labeled/assigned datasets

Kernel	MUTAG	ENZYMES	PTC_MR	PROTEINS	AIDS	MSRC_21C
SP	0.219	2.812	0.339	13.04	2.834	1.344
GH	7.875	382.929	34.328	5247.866	915.848	30.161
RW	75.856	9034.529	162.043	$OUT - OF - MEM$	20346.321	2470.96
GS	0.906	22.984	1.027	46.678	6.45	24.08
NH	0.12	1.926	0.307	6.996	9.304	**0.417**
WL-OA	0.401	17.571	1.88	308.433	140.45	1.675
NSPD	0.328	3.797	0.51	13.202	3.969	2.647
VEGK	**0.104**	**1.078**	**0.151**	**4.031**	**1.688**	0.531

Table 3. Accuracy of the classification (± td) on the 6 classification datasets containing graphs with labeled/assigned nodes

Kernel	MUTAG	ENZYMES	PTC_MR	PROTEINS	AIDS	MSRC_21C
SP	87.18(±0.99)	**62.03(±0.74)**	65.43(±1.48)	**76.74(±0.56)**	**99.59(±0.03)**	**85.72(±0.67)**
GH	83.34(±0.29)	42.33(±1.11)	58.62(±1.5)	76.33(±0.44)	99.42(±0.04)	27.34(±1.23)
RW	66.49(±0.0)	16.67(±0.0)	55.82(±0.0)	$OUT - OF - MEM$	80.0(±0.0)	13.88(±0.0)
GS	76.97(±0.38)	28.82(±1.15)	57.18(±0.48)	71.99(±0.36)	80.23(±0.04)	17.73(±1.69)
NH	90.15(±0.86)	58.58(±0.53)	66.08(±0.95)	75.77(±0.26)	99.44(±0.02)	63.39(±1.16)
WL-OA	88.54(±0.75)	59.25(±1.08)	**66.79(±1.08)**	76.11(±0.37)	99.36(±0.05)	80.78(±0.86)
NSPD	85.97(±1.04)	44.78(±0.99)	61.11(±1.23)	75.26(±0.25)	97.7(±0.14)	82.61(±0.62)
SREGK	86.65(±0.89)	44.53(±0.9)	59.82(±1.15)	71.52(±0.21)	98.88(±0.09)	15.84(±0.94)
VEGK	**91.0(±0.5)**	58.43(±0.49)	63.02(±0.78)	73.46(±0.63)	98.29(±0.13)	68.06(±0.79)

Results on Unlabeled Graphs Table 4 shows the computation times of the kernel matrices on the 4 datasets containing unlabeled graphs. Table 5 illustrates the classification accuracy scores obtained for all considered methods on the 4 unlabeled graph datasets. We notice that our kernel is the fastest on the first 2 datasets. On the Syntheticnew and Synthie datasets, our kernel ranks 2nd just after the **NH** kernel. In terms of classification accuracy, it proved competitive compared to other kernels. We can observe that on the first two datasets, our kernel is the most accurate. Our approach outperforms the **SREGK** approach except on the last dataset. This demonstrates once again the effectiveness of our improvements.

Table 4. Average running time of the kernel computation on the 4 datasets containing unlabeled graphs

Kernel	IMDB-BINARY	IMDB-MULTI	SYNTHETICnew	Synthie
SP	2.676	2.032	11.479	12.676
GH	40.854	49.877	237.854	363.177
RW	372.026	325.518	7354.625	10993.335
GS	589.696	541.116	89.724	178.158
NH	3.897	4.217	**1.344**	**2.099**
WL-OA	21.948	37.028	5.0	35.067
NSPD	11.934	12.211	18.265	20.366
VEGK	**1.724**	**1.703**	4.052	3.573

Table 5. Accuracy of the classification (± std) on the 4 datasets containing unlabeled graphs

Kernel	IMDB-BINARY	IMDB-MULTI	SYNTHETICnew	Synthie
SP	59.88(±0.19)	41.53(±0.25)	67.0(±1.31)	55.25(±1.79)
GH	50.0(±0.0)	33.33(±0.0)	50.0(±0.0)	27.5(±0.0)
RW	61.91(±0.44)	42.74(±0.11)	59.17(±0.84)	27.5(±0.0)
GS	66.99(±0.54)	43.21(±0.36)	66.63(±0.71)	54.25(±0.0)
NH	74.85(±0.39)	51.21(±0.55)	64.33(±0.0)	54.25(±0.0)
WL-OA	74.38(±0.52)	50.95(±0.34)	99.0(±0.27)	54.25(±0.0)
NSPD	73.38(±0.89)	52.23(±0.34)	**100.0(±0.0)**	54.1(±0.13)
SREGK	72.72(±0.53)	48.86(±0.5)	91.57(±0.52)	**58.73(±1.08)**
VEGK	**77.34(±0.74)**	**52.28(±0.5)**	98.4(±0.28)	54.35(0.14)

5   Conclusion

In this paper, we proposed a new graph kernel based on the concept of entropy. Our kernel, called VEGK (Von-Neumann Entropy Graph Kernel), brings an important improvement over existing entropy based kernels. VEGK uses Von Neumann entropy, which takes into account the labels/attributes of the graphs. Tests and experiments conducted on real-world datasets demonstrate the relevance of our kernel, which proved competitive compared to various graph kernels proposed in the literature.

Acknowledgments. This work was funded by Agence Nationale de la Recherche (ANR) under grant ANR-20-CE39-0008 and Département INFO-BOURG IUT Lyon1, campus de Bourg en Bresse.

References

1. Anand, K., Bianconi, G.: Entropy measures for networks: toward an information theory of complex topologies. Phys. Rev. E **80**(4), 045102 (2009)
2. Anand, K., Bianconi, G., Severini, S.: Shannon and Von Neumann entropy of random networks with heterogeneous expected degree. Phys. Rev. E **83**(3), 036109 (2011)
3. Borgwardt, K., Ghisu, E., Llinares-López, F., O'Bray, L., Rieck, B., et al.: Graph kernels: state-of-the-art and future challenges. Found. Trends® Mach. Learn. **13**(5–6), 531–712 (2020)
4. Borgwardt, K.M., Kriegel, H.P.: Shortest-path kernels on graphs. In: Fifth IEEE International Conference on Data Mining (ICDM 2005), pp. 8-pp. IEEE (2005)
5. Costa, F., De Grave, K.: Fast neighborhood subgraph pairwise distance kernel. In: ICML (2010)
6. Dairyko, M., et al.: Note on Von Neumann and Rényi entropies of a graph. Linear Algebra Appl. **521**, 240–253 (2017)
7. Feragen, A., Kasenburg, N., Petersen, J., de Bruijne, M., Borgwardt, K.: Scalable kernels for graphs with continuous attributes. In: Advances in Neural Information Processing Systems, vol. 26 (2013)
8. Hido, S., Kashima, H.: A linear-time graph kernel. In: 2009 Ninth IEEE International Conference on Data Mining, pp. 179–188. IEEE (2009)
9. Kang, U., Tong, H., Sun, J.: Fast random walk graph kernel. In: Proceedings of the 2012 SIAM International Conference on Data Mining, pp. 828–838. SIAM (2012)
10. Kersting, K., Kriege, N.M., Morris, C., Mutzel, P., Neumann, M.: Benchmark data sets for graph kernels (2016). http://graphkernels.cs.tu-dortmund.de
11. Kriege, N.M., Johansson, F.D., Morris, C.: A survey on graph kernels. Appl. Netw. Sci. **5**(1), 1–42 (2020)
12. Minello, G., Rossi, L., Torsello, A.: On the Von Neumann entropy of graphs. J. Complex Netw. **7**(4), 491–514 (2019)
13. Nikolentzos, G., Meladianos, P., Vazirgiannis, M.: Matching node embeddings for graph similarity. In: Proceedings of the 31th AAAI Conference on Artificial Intelligence, pp. 2429–2435, AAAI 2017 (2017)
14. Nikolentzos, G., Siglidis, G., Vazirgiannis, M.: Graph kernels: a survey. J. Artif. Intell. Res. **72**, 943–1027 (2021)

15. Nikolentzos, G., Vazirgiannis, M.: Enhancing graph kernels via successive embeddings. In: Proceedings of the 27th ACM International Conference on Information and Knowledge Management, pp. 1583–1586, CIKM 2018 (2018)
16. Salim, A., Shiju, S., Sumitra, S.: Graph kernels based on optimal node assignment. Knowl.-Based Syst. **244**, 108519 (2022)
17. Schulz, T.H., Horváth, T., Welke, P., Wrobel, S.: A generalized Weisfeiler-Lehman graph kernel. Mach. Learn. **111**, 2601–2629 (2022)
18. Shervashidze, N., Schweitzer, P., Van Leeuwen, E.J., Mehlhorn, K., Borgwardt, K.M.: Weisfeiler-Lehman graph kernels. J. Mach. Learn. Res. **12**(9), 2539–2561 (2011)
19. Shervashidze, N., Vishwanathan, S., Petri, T., Mehlhorn, K., Borgwardt, K.: Efficient graphlet kernels for large graph comparison. In: Artificial Intelligence and Statistics, pp. 488–495. PMLR (2009)
20. Togninalli, M., Ghisu, E., Llinares-López, F., Rieck, B., Borgwardt, K.: Wasserstein Weisfeiler-Lehman graph kernels. In: Advances in Neural Information Processing Systems, vol. 32 (2019)
21. Xu, L., Bai, L., Jiang, X., Tan, M., Zhang, D., Luo, B.: Deep Rényi entropy graph kernel. Pattern Recogn. **111**, 107668 (2021)
22. Xu, L., Jiang, X., Bai, L., Xiao, J., Luo, B.: A hybrid reproducing graph kernel based on information entropy. Pattern Recogn. **73**, 89–98 (2018)

Graph Neural Networks

GNN-DES: A New End-to-End Dynamic Ensemble Selection Method Based on Multi-label Graph Neural Network

Mariana de Araujo Souza[1]([✉]), Robert Sabourin[1],
George Darmiton da Cunha Cavalcanti[2], and Rafael Menelau Oliveira e Cruz[1]

[1] École de Technologie Supérieure, Université du Québec, Montréal QC, Canada
mariana.araujo.souza@gmail.com,
{robert.sabourin,rafael.menelau-cruz}@etsmtl.ca
[2] Centro de Informática, Universidade Federal de Pernambuco, Recife,
Pernambuco, Brazil
gdcc@cin.ufpe.br

Abstract. Most dynamic ensemble selection (DES) techniques rely solely on local information to single out the most competent classifiers. However, data sparsity and class overlap may hinder the region definition step, yielding an unreliable local context for performing the selection task. Thus, we propose in this work a DES technique that uses both the local information and classifiers' interactions to learn the ensemble combination rule. To that end, we encode the local information into a graph structure and the classifiers' information into multiple meta-labels, and learn the DES technique end-to-end using a multi-label graph neural network (GNN). Experimental results over 35 high-dimensional problems show the proposed method outperforms most evaluated DES techniques as well as the static baseline, suggesting its suitability for dealing with sparse overlapped data.

Keywords: Dynamic ensemble selection · Graph neural networks · Meta-learning · Data sparsity

1 Introduction

Dynamic ensemble selection (DES) techniques assume the classifiers in an ensemble make distinct mistakes in different areas of the feature space. Thus, they

---

The authors would like to thank the Canadian agencies FRQ (Fonds de Recherche du Québec) and NSERC (Natural Sciences and Engineering Research Council of Canada), and the Brazilian agencies CAPES (Coordenação de Aperfeiçoamento de Pessoal de Nível Superior), CNPq (Conselho Nacional de Desenvolvimento Científico e Tecnológico) and FACEPE (Fundação de Amparo à Ciência e Tecnologia de Pernambuco).

Supplementary Information The online version contains supplementary material available at https://doi.org/10.1007/978-3-031-42795-4_6.

M. Vento et al. (Eds.): GbRPR 2023, LNCS 14121, pp. 59–69, 2023.
https://doi.org/10.1007/978-3-031-42795-4_6

attempt to choose a subset of the models according to their perceived competence for classifying each sample in particular, often resulting in superior performances compared to static selection schemes, which label all test instances with the same set of classifiers [5]. Most DES techniques rely on the locality assumption to solve the dynamic selection task, in the sense that similar samples should be correctly labeled by a similar set of classifiers. These techniques require delimiting a region called the Region of Competence (RoC), via clustering [22], nearest neighbors rule [2,12,23], distance-based potential function [29], recursive partitioning [25], and/or fuzzy hyperboxes [6], in which the classifiers competences are estimated according to some criteria, such as local accuracy [12], classifier behavior [2], ensemble diversity [22], and meta-learning [3], among others.

The local region can thus have a large impact on the performance of these techniques [5], and so several methods attempt to directly improve its distribution. Filtering out the samples from the RoC is done in [17,18,24] based on the Item Response Theory (IRT) discrimination index, class distribution, and instance characterization, respectively. The RoC is characterized in [14] using a must link and a cannot link graph that are then used together with the classifiers' local accuracy to estimate their competence in the region. These techniques attempt to characterize and improve the local distribution for the classifier estimation step but they still rely solely on the locality assumption to compute a handcrafted competence estimation rule over an already defined region. While these approaches work generally well over a vast array of problems, such as class imbalanced distributions [17,24], local methods are known to struggle over high dimensionality and class ambiguity [26,31] and can present a strong sensitivity to overlap and data sparsity [21], with the latter being often associated with an increased class boundary complexity [10,15]. Such challenging scenarios may affect the local region definition and weaken the locality assumption, which in turn may limit the application of the dynamic selection techniques over real-world problems that present these characteristics, such as medical imaging data [7] and DNA microarray data [15] used for disease detection.

We also find in the literature a few dynamic selection techniques that do not rely on the locality assumption to perform the dynamic selection task [16,19]. Instead, they define the task as a multi-label meta-problem and learn the selection rule based on the classifiers' inter-dependencies, thus the meta-learner yields the ensemble combination rule for each input query instance without defining the RoC or explicitly estimating the classifiers' competences. While this approach could be interesting over the scenarios where the local context does not favor the dynamic classifier selection task, these techniques completely disregard the local context and can perform poorly against simple local accuracy-based techniques [19]. Both techniques also present a high computational cost due to the use of a meta-learner ensemble [19] and Monte Carlo sampling [16].

Thus, we propose in this work a dynamic selection technique that learns from the instances' relationships and classifiers' interactions jointly to better deal with high dimensional overlapped data. To that end, we model the data into a graph structure that can represent the samples' local and class inter-relations. We also model the classifiers' interactions as the multi-labels of the dynamic selection meta-problem. We then train a multi-label graph neural network (GNN) to yield the dynamic classifier combination rule in an end-to-end manner, without resorting to handcrafted meta-features or explicit RoC definition.

Graph neural networks operate directly on graph-structured data and are able to produce high-level representations of nodes and graphs [30]. The first GNNs were proposed for transductive learning and were unable to yield embeddings for unseen nodes, such as the Graph Convolutional Network (GCN) [11] which first generalized the convolution operation to the vertex domain. However, several models have been since proposed that work in inductive scenarios. The GraphSAGE model [9], seen as an extension of the GCN for inductive learning, learns a set of functions that aggregate the features from sampled neighboring nodes to produce the node embeddings. The Graph Attention Network (GAT) [28], which also works for inductive problems, presents a self-attention mechanism that allows the assignment of different weights to the neighbors in order to increase the model's capacity and to naturally deal with graphs that present variable node degrees.

Thus, by using a multi-label GNN as our meta-classifier, we leverage both the classifiers' inter-dependencies, represented in the meta-labels, and the samples' local interactions, represented in the graph, so that internally the network may learn an embedded space where the locality assumption for the dynamic selection task is stronger. We then contribute to the dynamic ensemble selection research area by (a) proposing an end-to-end technique that combines the information from the local data and the classifiers' interactions to better deal with sparse and overlapped data, and (b) evaluating the proposed method and ten other techniques over 35 high dimensional small sample sized (HDSSS) problems to assess whether learning from the two sources of information help overcome the limitations the current dynamic selection techniques present.

This work is organized as follows. The proposed method is introduced in Sect. 2. The experiments are reported in Sect. 3. Lastly, we summarize our conclusions in Sect. 4.

2   Graph Neural Network Dynamic Ensemble Selection Technique

We propose in this work the Graph Neural Network Dynamic Ensemble Selection (GNN-DES) technique, which attempts to better deal with locally complex scenarios in sparse overlapped data by combining the information from the samples' local context and the classifiers' interactions. To that end, we model the former using a graph structure, which is capable of representing the samples' local and class relationships, and model the latter by learning the dynamic selection task as a multi-label meta-problem.

Figure 1 describes the general steps of the GNN-DES technique. In memorization, the training set $\mathcal{T}$ and the pool of classifiers C are used to assign the samples' meta-labels U and construct the known graph $G_{\mathcal{T}}$, which are both then used to train the multi-label meta-learner GNN. In generalization, the query

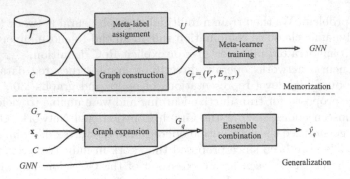

Fig. 1. Description of the GNN-DES technique. $\mathcal{T} = \{(\mathbf{x}_1, y_1), (\mathbf{x}_2, y_2), ..., (\mathbf{x}_N, y_N)\}$ is the training set and $U = \{\mathbf{u}_1, \mathbf{u}_2, ..., \mathbf{u}_N\}$ their corresponding meta-labels, $C = \{c_1, c_2, ..., c_{|C|}\}$ is the ensemble of classifiers. $G_\mathcal{T}$ is the known graph, composed of the set of training vertexes ($V_\mathcal{T}$) and edges ($E_{\mathcal{T} \times \mathcal{T}}$), G_q is the evaluation graph, composed of the $G_\mathcal{T}$ in addition to the query vertex (v_q) and its edges ($E_{q \times \mathcal{T}}$), and $\mathbf{x}_q$ and $\hat{y}_q$ are the query instance and its predicted label, respectively.

instance $\mathbf{x}_q$ is added to the known graph to produce the evaluation graph G_q, which is input to the meta-learner and used to produce the dynamic ensemble combination and then the output prediction $\hat{y}_q$.

Meta-label Assignment. In the meta-label assignment step, we characterize the competences of the classifiers by assigning to the samples meta-labels associated with their correct classification. This allows the meta-classifier to exploit the diverse behavior of the classifiers through learning the inter-dependecies between the meta-labels. Thus, to obtain the meta-labels, we evaluate the training set over the ensemble and we assign to each sample $(\mathbf{x}_i, y_i) \in \mathcal{T}$ a meta-label vector $\mathbf{u}_i$ of size $|C|$ so that $u_{i,k} = 1$ if the classifier c_k correctly labels $\mathbf{x}_i$, otherwise, $u_{i,k} = 0$.

Graph Construction. In the graph construction step, we aim to characterize in the known graph $G_\mathcal{T}$ the local context of the data that may be useful for the dynamic selection task. More specifically, we wish to embed the information of how reliable a sample is to indicate a good set of competent classifiers for another sample according to the locality assumption and the class relations. Thus, we link the samples that have a similar output response from the classifiers, as that may indicate they share a subset of competent classifiers. However, if the two samples belong to the same class we build a *strong* link, where the closer the samples the larger the edge weight as we expect the locality assumption to be stronger. Samples from different classes, on the other hand, are assigned a *weak* link, where the closer the samples the smaller the edge weight as the class ambiguity may indicate a weaker locality assumption.

Thus, to build the known graph $G_\mathcal{T}$, we project the training samples into the *decision* space, in which the axes represent the responses of each classifier in the pool. Then, we link each sample so that it has at least one strong link, to its nearest neighbor from the same class, and calculate its maximum margin for connection as a function of this link. Then, all samples within an instance's maximum margin are connected and their weights are set according to Eq. (1), where $d_{i,j}$ is the normalized $L1$ distance between the samples $(\mathbf{x}_i, y_i), (\mathbf{x}_j, y_j) \in \mathcal{T}$ projected into the decision space, d_i^{max} is the maximum margin for connection of $(\mathbf{x}_i, y_i)$, and τ is a preset threshold.

$$e_{i,j} = \begin{cases} 1 - d_{i,j}, \textit{if } (d_{i,j} \leq d_i^{max} \vee d_{i,j} \leq d_j^{max}) \wedge y_i = y_j, \\ d_{i,j}^2, \textit{if } (d_{i,j} \leq d_i^{max} \vee d_{i,j} \leq d_j^{max}) \wedge y_i \neq y_j, \\ 0, \textit{otherwise}, \end{cases}$$

$$d_i^{max} = min(d_{i,k}, \forall \mathbf{x}_k \in \mathcal{T} | y_k = y_i) + \tau \tag{1}$$

Meta-learner Training. Using the known graph $G_{\mathcal{T}}$ and the meta-labels U we fit the meta-learner in a supervised manner in the final step of the proposed method in memorization. We use a graph neural network core to learn and produce the node embeddings and a dense layer of size $|C|$ with sigmoid activation as the output layer of the network so that each output node of the network represents a classifiers' weight in the dynamic ensemble combination rule. We use a GNN core that is capable of inductive learning, and we fit the model using the binary cross-entropy loss, weighted so that the harder to classify the sample, the higher its weight, so as to encourage the model to focus on the more difficult samples. We measure the instance hardness as the number of classifiers in the pool that can label it correctly.

Graph Expansion. In generalization, we first expand the known graph to include the query instance in the data structure as to provide the meta-learner with its local context to obtain its ensemble combination rule. Thus, we project the query $\mathbf{x}_q$ into the decision space using the ensemble C and connect it to its nearest neighbor. Based on that, its maximum margin for connection is calculated and the edge weights between the query and the instances that fall within the margin are calculated as shown in (2), where $d_{q,j}$ is the normalized $L1$ distance between the samples $\mathbf{x}_q$ and $(\mathbf{x}_j, y_j) \in \mathcal{T}$ projected into the decision space, d_q^{max} is the query's maximum margin for connection, and τ is the preset threshold used in the graph construction step. The evaluation graph G_q is then built as the union between the known graph $G_{\mathcal{T}}$, the query vertex v_q, and the set of all its edges $E_q = \{e_{q,j}, \forall \mathbf{x}_j \in \mathcal{T}\}$.

$$e_{q,j} = \begin{cases} 1 - d_{q,j}, & \textit{if } d_{q,j} \leq d_q^{max}, \\ 0, & \textit{otherwise}, \end{cases}$$

$$d_q^{max} = min(d_{q,k}, \forall \mathbf{x}_k \in \mathcal{T}) + \tau \tag{2}$$

Ensemble Combination. We then induce the meta-learner GNN with the evaluation graph G_q to produce the network's outputs $\{o_{q,k}, \forall k \in |C|\}$, which represent the weighted support of each classifier when aggregating their responses. The class with the largest support is output as the query's predicted label $\hat{y}_q$.

3 Experiments

We evaluate in the experiments how well the DES techniques perform against a static selection baseline, and whether the proposed method is able to outperform them over the HDSSS problems. We describe the experimental protocol and present the results next.

Ensemble Methods. We use as our baseline an AdaBoost (ADA) [8] ensemble composed of 100 Decision Stumps, and we also evaluate 10 dynamic ensemble methods, namely: the K-Nearest Oracles Union (KNU) [12], the K-Nearest Oracles Eliminate (KNE) [12], the Dynamic Ensemble Selection-KNN (DKNN) [22], the K-Nearest Output Profiles (KNOP) [2], the META-DES [3], the Randomized Reference Classifier (RRC) [29], the Chained Dynamic Ensemble (CHADE) [19], the Online Local Pool technique (OLP) [23], the OLP++ [25] and the Forest of Local Trees (FLT) [1]. Except for CHADE, all of them are local-based techniques, though they may define the RoC using distinct methods or in different spaces, and the OLP, OLP++ and FLT are not evaluated using the AdaBoost ensemble as they produce their own pool. We also include the performance of the Oracle [13], an abstract model that always selects the correct classifier if it exists, to provide an upper limit to the performance of the DES techniques.

Hyperparameters. The techniques' hyperparameters were set as recommended in their papers if no implementation is available in the DESLib [4] library, or to their default value otherwise. The GNN-DES threshold was set to $\tau = 0.05$, and the meta-learner contained two GraphSAGE [9] layers of size 512 units, as in [20], and one dense output layer, as in [9]. We use the attentional aggregation function from [28] in the convolutional layers as the local samples may have distinct importances for the DES task. To cope with the small sample-sized problems, we sampled only 5 samples in each convolutional layer and applied L_2 regularization with $\lambda = 0.01$, which was empirically observed to help the training according to the validation loss curves. Moreover, 20% of the training set was used for validation/early stopping, and the validation nodes were connected to the known graph G_T as if unknown samples (2). The GNN was trained over 150 epochs, with a patience of 30 epochs, a batch size of 300, and the adaptive learning rate is initially set to 0.005, as in [28]. We also performed a sweep on the drop-out rate in the set $\{0.0, 0.2, 0.5\}$ as in [20], and the model with the best micro-averaged multi-label precision in validation was chosen.

Datasets and Evaluation. We use the datasets shown in Table 1, which are the same set of problems used in [25] with the exception of four datasets over which the ensemble method generated fewer than the set amount of classifiers in the pool. The testbed contains two-class HDSSS datasets (with at least 100 features) taken from the OpenML repository [27]. The columns N, F, and IR indicate the problems' number of instances, number of features and imbalance ratio, respectively, while the ratio F/N conveys the problems' sparsity and is associated with a higher data complexity [10,15].

We evaluate the datasets using a 10-fold cross-validation procedure using the folds available at the repository for reproducibility, and as in [25] we use the training set as the dynamic selection set (DSEL), a labeled dataset used for RoC definition [5], due to the limited number of instances in several datasets. Also due to the varying imbalance ratios in the testbed, we use the macro-averaged recall, or balanced accuracy rate, as the performance measure, to account for the class disproportion without focusing on one of the classes.

Table 1. Characteristics of the datasets used in the experiments.

Dataset	N	F	IR	F/N	Dataset	N	F	IR	F/N
tumors_C	60	7129	1.86	118.82	OVA_Endometrium	1545	10935	24.33	7.08
leukemia	72	7129	1.88	99.01	OVA_Uterus	1545	10935	11.46	7.08
AP_Endometrium_Lung	187	10935	2.07	58.48	OVA_Ovary	1545	10935	6.80	7.08
AP_Omentum_Uterus	201	10935	1.61	54.40	OVA_Breast	1545	10935	3.49	7.08
AP_Omentum_Lung	203	10935	1.64	53.87	fri_c4_100_100	100	100	1.13	1.00
AP_Lung_Uterus	250	10935	1.02	43.74	tecator	240	124	1.35	0.52
AP_Omentum_Ovary	275	10935	2.57	39.76	fri_c4_250_100	250	100	1.27	0.40
AP_Ovary_Uterus	322	10935	1.60	33.96	gina_agnostic	3468	970	1.03	0.28
AP_Omentum_Kidney	337	10935	3.38	32.45	gina_prior	3468	784	1.03	0.23
AP_Colon_Prostate	355	10935	4.14	30.80	fri_c4_500_100	500	100	1.30	0.20
AP_Colon_Omentum	363	10935	3.71	30.12	spectrometer	531	101	8.65	0.19
AP_Uterus_Kidney	384	10935	2.10	28.48	scene	2407	299	4.58	0.12
AP_Endometrium_Breast	405	10935	5.64	27.00	mfeat-pixel	2000	240	9.00	0.12
AP_Breast_Prostate	413	10935	4.99	26.48	mfeat-factors	2000	216	9.00	0.11
AP_Breast_Omentum	421	10935	4.47	25.97	fri_c4_1000_100	1000	100	1.29	0.10
AP_Colon_Ovary	484	10935	1.44	22.59	yeast_ml8	2417	116	70.09	0.05
AP_Colon_Kidney	546	10935	1.10	20.03	sylva_prior	14395	108	15.25	0.01
AP_Breast_Kidney	604	10935	1.32	18.10					

Results. Table 2 summarizes the performances of the Oracle, the static selection baseline (ADA) and the other 10 dynamic ensemble methods besides the proposed GNN-DES. The average performances per dataset are available in the supplementary material. We can see that the GNN-DES yielded the highest average balanced accuracy rate and the highest average rank among all techniques. Moreover, the GNN-DES obtained a higher average performance over at least half of the datasets compared to all techniques except for the META-DES, another local-based meta-learning technique.

Performing the non-parametric Wilcoxon signed-rank test over the pairs of techniques, we obtain the p-values shown in Table 3. First, we observe that the GNN-DES statistically outperformed with significance $\alpha = 0.05$ all evaluated techniques except the KNOP and the META-DES. As these three best-performing and statistically similar techniques are the only ones to rely on the local information in the decision space (in addition to the feature space, in the case of the META-DES), the results suggest that this approach may be better indicated for dynamic classifier selection on HDSSS problems.

We can also observe in Table 3 that the GNN-DES was the only dynamic ensemble method to statistically outperform the static selection baseline (ADA) with $\alpha = 0.05$ over the HDSSS datasets. This suggests that not only do the DES techniques generally struggle over these sparse datasets, as could be reasonably expected, but also that GNN-DES might behave somewhat differently from the classical local-based approaches, possibly due to the inclusion of the other source of information relative to the classifiers' interactions. However, how exactly the learned embedded space may affect the behavior of the GNN-DES and in which situations this information is valuable are questions to be analyzed in the future. All in all, we believe these promising results over the HDSSS problems warrant further investigation into the proposed approach.

Table 2. Mean balanced accuracy rate and rank, averaged over all datasets. The *Win-tie-loss* row refers to the number of datasets the GNN-DES obtained a higher, equal, or lower average performance to the column-wise technique.

	Oracle	ADA	KNU	KNE	DKNN	KNOP	META-DES	RRC	CHADE	FLT	OLP	OLP++	GNN-DES
Mean	99.97	88.14	87.12	84.08	84.73	88.88	88.90	85.30	83.93	84.88	81.35	86.22	**89.03**
Mean rank	n/a	5.31	5.36	8.39	8.01	4.90	4.29	6.33	7.69	5.90	10.59	7.11	**4.13**
Win-tie-loss	n/a	23-0-12	21-3-11	31-0-4	32-0-3	18-1-16	15-0-20	21-3-11	28-0-7	22-1-12	33-0-2	27-1-7	n/a

Table 3. Resulting p-values of the Wilcoxon signed-rank test between average balanced accuracy rates of all pairs of techniques, rounded to the second decimal point. Values below $\alpha = 0.05$ are in bold, rounded values below 0.01 are underlined, and the symbols $\pm$ indicate whether the column-wise technique statistically outperformed or not the row-wise technique.

	ADA	KNU	KNE	DKNN	KNOP	META-DES	RRC	CHADE	FLT	OLP	OLP++	GNN-DES
ADA	n/a	0.38	**0.01**(−)	**0.01**(−)	0.10	0.10	0.22	**0.01**(−)	0.51	**0.01**(−)	**0.01**(−)	**0.04**(+)
KNU		n/a	**0.01**(−)	**0.01**(−)	0.11	0.08	0.06	**0.01**(−)	0.86	**0.01**(−)	0.37	**0.03**(+)
KNE			n/a	0.26	**0.01**(+)	**0.01**(+)	0.24	0.66	0.15	**0.01**(−)	0.15	**0.01**(+)
DKNN				n/a	**0.01**(+)	**0.01**(+)	0.42	0.95	0.30	**0.01**(−)	0.21	**0.01**(+)
KNOP					n/a	0.12	0.06	**0.01**(−)	0.65	**0.01**(−)	**0.01**(−)	0.62
META-DES						n/a	**0.02**(−)	**0.01**(−)	0.27	**0.01**(−)	**0.01**(−)	0.59
RRC							n/a	0.12	0.67	**0.01**(−)	0.85	**0.01**(+)
CHADE								n/a	0.35	0.09	0.25	**0.01**(+)
FLT									n/a	**0.01**(−)	0.47	0.13
OLP										n/a	**0.01**(+)	**0.01**(+)
OLP++											n/a	**0.01**(+)
GNN-DES												n/a

4 Conclusion

We proposed in this work the GNN-DES technique, which learns the dynamic classifier combination rule from the instances' relationships and classifiers' interactions to deal with sparse overlapped data. We encode the local and class relations between the samples into a graph structure and the ensemble competence information into multiple meta-labels, and then fit our meta-learner, a multi-label GNN model, to perform the DES task in an end-to-end manner.

Experiments over 35 HDSSS datasets showed that the DES techniques in the literature had difficulty in surpassing the static selection baseline, especially the techniques based solely on similarities in the feature space for RoC definition. The locality assumption in the decision space was shown to perform better over the sparse data, and the three techniques that use this approach performed similarly and the best. Moreover, the GNN-DES was the only technique to statistically outperform the baseline in addition to 8 of the 10 evaluated DES techniques, suggesting its suitability for dealing with sparse and overlapped data.

Future work in this line of research may involve evaluating the impact of using different ensemble methods and hyperparameters to analyze the relationship between the graph characteristics and the technique's performance. Furthermore, we may analyze the behavior of the technique in different local contexts and its relation to the learned embedded space to investigate in which scenarios the meta-learner improves the locality assumption for the DES task.

References

1. Armano, G., Tamponi, E.: Building forests of local trees. Pattern Recogn. **76**, 380–390 (2018)
2. Cavalin, P.R., Sabourin, R., Suen, C.Y.: LoGID: an adaptive framework combining local and global incremental learning for dynamic selection of ensembles of HMMs. Pattern Recogn. **45**(9), 3544–3556 (2012)
3. Cruz, R.M.O., Sabourin, R., Cavalcanti, G.D.C., Ren, T.I.: META-DES: a dynamic ensemble selection framework using meta-learning. Pattern Recogn. **48**(5), 1925–1935 (2015)
4. Cruz, R.M.O., Hafemann, L.G., Sabourin, R., Cavalcanti, G.D.C.: DESlib: a dynamic ensemble selection library in python. J. Mach. Learn. Res. **21**(8), 1–5 (2020)
5. Cruz, R.M., Sabourin, R., Cavalcanti, G.D.: Dynamic classifier selection: recent advances and perspectives. Inf. Fusion **41**, 195–216 (2018)
6. Davtalab, R., Cruz, R.M., Sabourin, R.: Dynamic ensemble selection using fuzzy hyperboxes. In: 2022 International Joint Conference on Neural Networks (IJCNN), pp. 1–9 (2022)
7. El-Sappagh, S., et al.: Alzheimer's disease progression detection model based on an early fusion of cost-effective multimodal data. Futur. Gener. Comput. Syst. **115**, 680–699 (2021)
8. Freund, Y., Schapire, R.E.: A decision-theoretic generalization of on-line learning and an application to boosting. J. Comput. Syst. Sci. **55**, 119–139 (1997)

9. Hamilton, W., Ying, Z., Leskovec, J.: Inductive representation learning on large graphs. In: Advances in Neural Information Processing Systems, pp. 1024–1034 (2017)
10. Ho, T.K., Basu, M.: Complexity measures of supervised classification problems. IEEE Trans. Pattern Anal. Mach. Intell. **24**(3), 289–300 (2002)
11. Kipf, T.N., Welling, M.: Semi-supervised classification with graph convolutional networks. In: International Conference on Learning Representations (ICLR) (2017)
12. Ko, A.H.-R., Sabourin, R., de Souza Britto Jr., A.: A new dynamic ensemble selection method for numeral recognition. In: Haindl, M., Kittler, J., Roli, F. (eds.) MCS 2007. LNCS, vol. 4472, pp. 431–439. Springer, Heidelberg (2007). https://doi.org/10.1007/978-3-540-72523-7_43
13. Kuncheva, L.I.: A theoretical study on six classifier fusion strategies. IEEE Trans. Pattern Anal. Mach. Intell. **24**(2), 281–286 (2002)
14. Li, D., Wen, G., Li, X., Cai, X.: Graph-based dynamic ensemble pruning for facial expression recognition. Appl. Intell. **49**(9), 3188–3206 (2019)
15. Lorena, A.C., Costa, I.G., Spolaôr, N., De Souto, M.C.: Analysis of complexity indices for classification problems: cancer gene expression data. Neurocomputing **75**(1), 33–42 (2012)
16. Narassiguin, A., Elghazel, H., Aussem, A.: Dynamic ensemble selection with probabilistic classifier chains. In: Ceci, M., Hollmén, J., Todorovski, L., Vens, C., Džeroski, S. (eds.) ECML PKDD 2017. LNCS (LNAI), vol. 10534, pp. 169–186. Springer, Cham (2017). https://doi.org/10.1007/978-3-319-71249-9_11
17. Oliveira, D.V., Cavalcanti, G.D., Porpino, T.N., Cruz, R.M., Sabourin, R.: K-nearest oracles borderline dynamic classifier ensemble selection. In: 2018 International Joint Conference on Neural Networks (IJCNN), pp. 1–8. IEEE (2018)
18. Pereira, M., Britto, A., Oliveira, L., Sabourin, R.: Dynamic ensemble selection by K-nearest local Oracles with discrimination index. In: 2018 IEEE 30th International Conference on Tools with Artificial Intelligence, pp. 765–771. IEEE (2018)
19. Pinto, F., Soares, C., Mendes-Moreira, J.: CHADE: metalearning with classifier chains for dynamic combination of classifiers. In: Frasconi, P., Landwehr, N., Manco, G., Vreeken, J. (eds.) ECML PKDD 2016. LNCS (LNAI), vol. 9851, pp. 410–425. Springer, Cham (2016). https://doi.org/10.1007/978-3-319-46128-1_26
20. Salehi, A., Davulcu, H.: Graph attention auto-encoders. In: 2020 IEEE 32nd International Conference on Tools with Artificial Intelligence, pp. 989–996 (2020)
21. Sánchez, J.S., Mollineda, R.A., Sotoca, J.M.: An analysis of how training data complexity affects the nearest neighbor classifiers. Pattern Anal. Appl. **10**(3), 189–201 (2007)
22. Soares, R.G., Santana, A., Canuto, A.M., de Souto, M.C.P.: Using accuracy and diversity to select classifiers to build ensembles. In: The 2006 IEEE International Joint Conference on Neural Network (IJCNN) Proceedings, pp. 1310–1316 (2006)
23. Souza, M.A., Cavalcanti, G.D., Cruz, R.M., Sabourin, R.: Online local pool generation for dynamic classifier selection. Pattern Recogn. **85**, 132–148 (2019)
24. Souza, M.A., Sabourin, R., Cavalcanti, G.D.C., Cruz, R.M.O.: Local overlap reduction procedure for dynamic ensemble selection. In: 2022 International Joint Conference on Neural Networks (IJCNN), pp. 1–9 (2022)
25. Souza, M.A., Sabourin, R., Cavalcanti, G.D., Cruz, R.M.: OLP++: an online local classifier for high dimensional data. Inf. Fusion **90**, 120–137 (2023)
26. Vandaele, R., Kang, B., De Bie, T., Saeys, Y.: The curse revisited: when are distances informative for the ground truth in noisy high-dimensional data? In: International Conference on Artificial Intelligence and Statistics, pp. 2158–2172. PMLR (2022)

27. Vanschoren, J., van Rijn, J.N., Bischl, B., Torgo, L.: OpenML: networked science in machine learning. SIGKDD Explor. **15**(2), 49–60 (2013)
28. Veličković, P., Cucurull, G., Casanova, A., Romero, A., Liò, P., Bengio, Y.: Graph attention networks. In: International Conference on Learning Representations (2018)
29. Woloszynski, T., Kurzynski, M.: A probabilistic model of classifier competence for dynamic ensemble selection. Pattern Recogn. **44**(10), 2656–2668 (2011)
30. Xia, F., et al.: Graph learning: a survey. IEEE Trans. Artif. Intell. **2**(2), 109–127 (2021)
31. Zhang, S.: Challenges in KNN classification. IEEE Trans. Knowl. Data Eng. **34**(10), 4663–4675 (2022)

C2N-ABDP: Cluster-to-Node Attention-Based Differentiable Pooling

Rongji Ye[1], Lixin Cui[1]([✉]), Luca Rossi[2], Yue Wang[1], Zhuo Xu[1],
Lu Bai[3], and Edwin R. Hancock[1,4]

[1] Central University of Finance and Economic, Beijing, China
cuilixin@cufe.edu.cn
[2] Department of Electrical and Electronic Engineering,
The Hong Kong Polytechnic University, Hong Kong SAR, China
[3] School of Artificial Intelligence, Beijing Normal University, Beijing, China
[4] Department of Computer Science, University of York, York, UK

Abstract. Graph neural networks have achieved state-of-the-art performance in various graph based tasks, including classification and regression at both node and graph level. In the context of graph classification, graph pooling plays an important role in reducing the number of graph nodes and allowing the graph neural network to learn a hierarchical representation of the input graph. However, most graph pooling methods fail to effectively preserve graph structure information and node feature information when reducing the number of nodes. At the same time, the existing hierarchical differentiable graph pooling methods cannot effectively calculate the importance of nodes and thus cannot effectively aggregate node information. In this paper, we propose an attention-based differentiable pooling method, which aggregates nodes into clusters when reducing the scale of the graph, uses singular value decomposition to calculate cluster information during the aggregation process, and captures node importance information through a novel attention mechanism. The experimental results show that our approach outperforms competitive models on benchmark datasets.

Keywords: Graph neural network · Graph pooling · Attention

1 Introduction

The success of Convolutional Neural Networks (CNNs) [3,4,14,17,18], which were able to achieve state-of-the-art performance on image classification tasks, inspired many researchers to try and generalize CNNs from the image domain to the graph one [2,6,8,9]. However, there are important differences between graphs and images. Crucially, images are regular and grid-like, while graphs can have highly complex, irregular structure. This in turn makes it difficult to directly apply popular CNN architectures to tasks that involve the analysis of graphs.

For this reason, in recent years there has been an increasing interest in extending existing neural architectures, such as CNNs, to the graph domain. In this context, several Graph Neural Networks (GNNs) models have been proposed to explore the information contained in node and edge features, while also capturing the structural information of graphs. The GCN model [15] utilizes the Fourier transform and the graph Laplacian matrix to define a convolution operation on graphs. GraphSAGE [13] optimizes the full graph sampling of GCNs to partial node-centric neighbor sampling, making distributed training of large-scale graph data possible. The GAT architecture [28] combines the attention mechanism with the original GCN model, alleviating the bottleneck problem of GCNs and allowing the assignment of different attention coefficients to different neighbors. With the introduction of DGCNNs [33], the authors proposed an end-to-end graph neural network architecture that uses SortPooling to classify vertex features instead of summing them. Another well-known model is GIN [30], which uses an injective aggregation strategy, and its ability to identify graph structures is comparable to Weisfeiler-Lehman (WL) tests. These GNNs have been widely employed for both node and graph level classification tasks.

In this paper, we focus on the graph classification task, i.e., predicting the class of whole graphs. For example, chemical molecules can be represented as graphs. In this case, we can extract graph node features and graph structure information, learn a low-dimensional representation vector corresponding to the graph data, and then use this to predict a molecular property of interest. At present, graph representation learning methods can achieve excellent results in graph classification tasks. Often, these models use some form of graph pooling.

Graph pooling methods can be divided into global and hierarchical pooling methods [32]. Global pooling methods, such as mean and max pooling, aggregate all nodes equally during pooling and thus lead to loss of structural information and neglect the differences among the nodes. Hierarchical graph pooling methods, on the other hand, extract the local subgraph structure information and gradually aggregate the subgraph information to obtain a representation vector for the full graph. The subgraph structure may play a very important role in the classification of the whole graph. For example, the categories of organic molecules can be divided according to the functional groups in the molecules.

In recent years, many effective graph pooling methods have emerged. DIFFPool [32] is a differentiable graph pooling method that assigns nodes to clusters by learning a soft clustering node assignment matrix. The node representations are weighted and summed according to the probability of the nodes belonging to the cluster, and then the cluster representation vectors are obtained. The cluster representation vectors obtained by the pooling layer are then used as the new node representations in the next layer. TOPK [12] is a method that discards nodes in each pooling process to reduce the size of the graph. Compared with DIFFPool, it has the advantage of reducing the number of parameters and improving the computational efficiency, however this comes at the cost of a lower performance. SAGPool [19] is a pooling method combined with an attention mechanism. Originally based on TOPK, SAGPool uses graph convolution to achieve self-attention, obtain the importance score of each node, and retain the nodes with the top k highest scores. Compared with TOPK, SAGPool considers the local structure information of the graph when scoring the nodes. The main drawback, however, is that the structure of the pooled graph is obtained through a simple deletion process on the original graph. HGP-SL [34] uses the Manhattan distance between node representations and the representations constructed by their neighboring nodes as the criteria for filtering nodes. Then, it retains those nodes that are difficult to be reconstructed by their neighbors. HGP-SL also uses an attention-based structure learning layer to learn the potential connections between nodes. VIPool [20] calculates the mutual information between a node and its neighbors to measure the similarity and difference between each node and its neighbors, and then retains nodes that can reflect the information of their neighboring nodes. ASAP [23] takes each node as the center to obtain candidate clusters, utilizes a self-attention mechanism to learn cluster representation and a novel way to filter candidate clusters and complete the pooling process.

In this work, we propose Cluster-to-Node Attention-based Differentiable Pooling (C2N-ABDP), a new attention-based graph pooling method based on DIFFPool that is able to effectively utilize the graph structure information and that is suitable for hierarchical pooling architectures. The main contributions of our work can be summarized as follows: 1) we propose a Cluster-to-Node attention (C2N attention) mechanism and we combine it with DiffPool, allowing the model to effectively learn the importance of nodes within clusters and thus to obtain more reasonable cluster representations; 2) we apply SVD to extract the cluster information in a way that is robust to the order of nodes in the cluster.

The remainder of the paper is organized as follows. In Sect. 2 we discuss the relevant background and the limitations of two close methods to ours, DIFF-Pool and ABDP [21]. In Sect. 3 we introduce C2N-ABDP and we present the experimental results in Sect. 4. Finally, Sect. 5 concludes the paper.

2   Background

2.1   Graph Convolutional Networks

The GCN model [13] stems from graph signal theory and uses the Fourier transform as a bridge to successfully map CNNs to the graph domain and is now widely used for feature extraction from graphs. Given a graph with adjacency matrix A and node features matrix X, GCNs work by computing a new matrix of convolved features Z, i.e.,

$$Z = \tilde{D}^{-1/2} \tilde{A} \tilde{D}^{-1/2} X \Theta, \tag{1}$$

where $\tilde{A} = A + I_N$, I_N is the graph adjacency matrix with self-connections added, I_N is the unit diagonal matrix, $\tilde{D}$ is the degree matrix of $\tilde{A}$, and Θ is a trainable parameter matrix. GCNs aggregate the features of each node and its neighbor nodes to obtain a new node feature and they are widely used in hierarchical pooling architectures for graph feature extraction.

2.2 The Attention Mechanism

The attention mechanism [27] is one of the most widely used components in the field of deep learning for natural language processing. The attention mechanism calculates three matrices based on the source data matrix $H \in \mathbb{R}^{n \times d}$, i.e.,

$$Q = HW_q, \ K = HW_k, \ V = HW_v \tag{2}$$

$$\text{Attention}(Q, K, V) = \text{Softmax}\left(\frac{QK^\top}{\sqrt{d_k}}\right) V, \tag{3}$$

where $Q \in \mathbb{R}^{n \times d_q}$ and $K \in \mathbb{R}^{n \times d_q}$ (the query and key matrices, respectively) are used to calculate the attention score $\text{Attention}(Q, K, V)$, $V \in \mathbb{R}^{n \times d_v}$ is the representation matrix, d_q is the dimension of attention calculation, d_v is the feature dimension of the output, and the softmax operates on each row of $QK^\top$ to get the importance score.

2.3 Graph Pooling

DIFFPool is a differentiable pooling method that learns a soft assignment matrix and uses it to generate a new adjacency matrix and new node representation vectors. Each value in the soft assignment matrix represents the probability that a given node belongs to a corresponding cluster. This allows DIFFPool to effectively utilize graph structure information, but it does not sufficiently evaluate the importance of nodes within each cluster. In particular, we argue that the importance of a node within a cluster is not necessarily equivalent to the probability that the node belongs to the cluster.

ABDP [21] combines DIFFPool with an attention mechanism. It uses the learned soft assignment matrix to generate a hard assignment matrix and ensure that a node can only be assigned to one cluster. Then, it calculates the mutual attention between nodes belonging to the same cluster. The problem with ABDP is that the meaning of the attention mechanism is not clear and it lacks the ability to extract overall information on the cluster.

2.4 Singular Value Decomposition

Singular value decomposition (SVD) is often used in data dimensionality reduction and recommendation systems. Given an input matrix $A \in \mathbb{R}^{m \times n}$, SVD computes the decomposition $A = U\Sigma V^\top$, with $U \in \mathbb{R}^{m \times m}$, $\Sigma \in \mathbb{R}^{m \times n}$ and $V \in \mathbb{R}^{n \times n}$. Both U and V are eigenvector matrices, U mainly contains information related to the rows of A and V mainly contains information related to the columns of A. Σ is the diagonal eigenvalue matrix, where each diagonal value of Σ represents the importance of the corresponding eigenvectors in U and V. We note that in general Σ and V are independent of the row order of A, which in turn ensures that SVD can be applied as part of our pooling method.

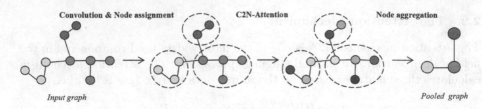

Convolution & Node assignment   C2N-Attention   Node aggregation

Input graph   Pooled graph

Fig. 1. Overview of the proposed pooling mechanism. C2N-ABDP collapses the graph through graph convolution, hard assignment matrix generation, cluster information extraction, and C2N attention embedding aggregation.

3   The Proposed Pooling Method

We propose a novel pooling operator, named Cluster-to-Node Attention-based Differentiable Pooling (C2N-ABDP). At layer l, C2N-ABDP performs 5 main operations: 1) graph convolution, 2) hard assignment matrix computation, 3) cluster information extraction, 4) C2N attention computation, and 5) adjacency matrix update and embedding matrix computation.

(i) Graph convolution. Given the adjacency matrix $A^{(l)}$ and the feature matrix $X^{(l)}$, we use a GCN to generate the embedding matrix $Z^{(l)} \in \mathbb{R}^{n_l \times d}$ according to Eq. 1.

(ii) Hard assignment matrix computation. We use another GCN to generate the soft assignment matrix $S^{(l)} \in \mathbb{R}^{n_l \times n_{l+1}}$, i.e.,

$$S^{(l)} = \text{Softmax}\left(\text{GCN}_{l,\text{pool}}\left(A^{(l)}, X^{(l)}\right)\right). \tag{4}$$

Each row of $S^{(l)}$ represents the probability of a node of the l-th layer belonging to a cluster of the $(l+1)$-th layer. Given $S^{(l)}$, we generate the hard assignment matrix $HS^{(l)} \in \mathbb{R}^{n_l \times n_{l+1}}$ by mapping the maximum value of each row in $S^{(l)}$ to 1 and the remaining values to 0, so that each node will only be assigned to a single cluster.

(iii) Cluster information extraction. The cluster information extraction procedure can be further subdivided into 2 main steps.

SVD process. With $HS^{(l)}$ to hand, we define the matrix $Z_{(t)}^{(l)} \in \mathbb{R}^{n_t \times d}$ of the node vectors assigned to t-th cluster, where t denotes the cluster number at layer l, n_t is the number of nodes assigned to cluster t, and d is the dimension of node features in the current layer. Then, SVD is used to decompose $Z_{(t)}^{(l)}$ into $U_{(t)}^{(l)} \in \mathbb{R}^{n_t \times n_t}$, $S_{(t)}^{(l)} \in \mathbb{R}^{n_t \times d}$, $V_{(t)}^{(l)} \in \mathbb{R}^{d \times d}$. Note that the order with which the nodes are allocated to a cluster by the hard assignment matrix is not fixed, however $S_{(t)}^{(l)}$ and $V_{(t)}^{(l)}$ are not affected by the order of nodes in the cluster. We extract the preliminary cluster information as

$$M_{(t)}^{(l)} = S_{(t)}^{(l)} V_{(t)}^{(l)} \in \mathbb{R}^{n_t \times d}. \tag{5}$$

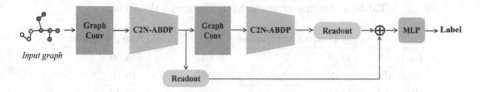

Fig. 2. Structure of the proposed architecture.

Our approach ensures that no matter how the nodes allocated to a cluster are ordered, as long as the node features are the same, the extracted cluster information is the same.

Residual linear layer. We first perform a zero padding operation on $M_{(t)}^{(l)}$ to expand its dimension to get $M_{(t)}^{'(l)} \in \mathbb{R}^{d \times d}$, with $d > n_t$. The cluster information is computed as

$$C_{(t)}^{(l)\top} = L0(M_{(t)}^{(l)} + L3(L2(L1(WM_{(t)}^{(l)})))), \tag{6}$$

where L0, L1, L2, and L3 are four linear layers, the output dimensions of L1 and L3 are both one, and $C_{(t)}^{(l)} \in \mathbb{R}^{1 \times d}$ is the extraction result of the cluster information, which will be used to determine the attention. The residual structure is introduced to prevent overfitting. Finally, note that the previous equation omits the activation function ReLU [1].

(iv) Cluster-to-node attention. The calculation of the C2N attention score follows the equations

$$Q = C_{(t)}^{(l)} W_q \in \mathbb{R}^{1 \times d_k}, \quad K = Z_{(t)}^{(l)} W_k \in \mathbb{R}^{n_t \times d_k} \tag{7}$$

$$AT_{(t)}^{(l)} = \text{Softmax}\left(\frac{QK^\top}{d_k}\right) \in \mathbb{R}^{1 \times n_t}. \tag{8}$$

The process is similar to the computation of the scaled dot-product attention [27], however here we changed the source of Q. Q is obtained from the cluster information matrix $C_{(t)}^{(l)}$ and K is obtained from $Z_{(t)}^{(l)}$, i.e., the representation matrix of the nodes in the cluster. $AT_{(t)}^{(l)}$ is the resulting attention vector, representing the importance of each node to the cluster, while W_q and W_k are trainable parameters.

(v) Adjacency matrix update and embedding matrix. We use the hard cluster assignment matrix and the adjacency matrix of the current layer to calculate the adjacency matrix of the next layer. The node feature matrix of the next layer is given by the concatention of the cluster representation vectors,

Table 1. Summary statistics of the graph datasets.

	# Graphs	# Classes	Avg. # Nodes	Avg. # Edges
D&D	1178	2	284.32	715.66
MUTAG	188	2	17.93	19.79
PROTEINS	1113	2	39.06	72.82
NCI1	4110	2	29.87	32.30
ENZYMES	600	6	32.63	62.14
PTC	344	2	14.29	14.69

and the cluster representation is the weighted sum of the row vectors of $Z_{(t)}^{(l)}$ wrt the attention vector $AT_{(t)}^{(l)}$, i.e.,

$$X^{(l+1)} = \|_{t=1}^{n_{l+1}} AT_{(t)}^{(l)} Z_{(t)}^{(l)} \in \mathbb{R}^{n_{l+1} \times d} \tag{9}$$

$$A^{(l+1)} = HS^{(l)\top} A^{(l)} HS^{(l)} \in \mathbb{R}^{n_{l+1} \times n_{l+1}}. \tag{10}$$

Figure 1 shows the graph collapsing process of C2N-ABDP. Nodes assigned to different clusters are marked with different colors, and nodes assigned to the same cluster by the hard assignment matrix are surrounded by dotted lines. The C2N attention is applied to calculate the importance of nodes in the cluster, where darker shades denote highlight more important nodes, e.g., within the cluster of blue nodes, darker blue nodes are more important than lighter blue nodes. Finally, according to the importance of the nodes, the representations of the nodes in the cluster are aggregated to obtain the next layer node representations.

4   Experiments

We evaluate the performance of a hierarchical pooling architecture based on C2N-ABDP. Figure 2 shows the overall structure of our architecture. The input graph data passes through two encoding blocks, each block contains a standard GCN and the proposed pooling operator C2N-ABDP. ⊕ represents the splicing operation, and after splicing the output of the two blocks, the representation of the graph is obtained. Finally, the representation of the graph is passed through the MLP classifier to obtain the classification result.

We compare our architecture with three graph kernel methods and nine GNNs. The three graph kernel methods are GK [26], WL [25], and DGK [31]. The nine GNN methods are DGCNN [33], GIN [30], SAGPool [19], DIFFPool [32], ABDP [21], ASAP [23], VIPool [20], MinCutPool [5], CGIPool [22]. We use results reported by the original papers (when available) in Table 2.

The models are compared in terms of classification accuracy on six graph classification benchmarks. These datasets are D&D [11], MUTAG [10], PRO-TEINS [7], NCL1 [29], ENZYMES [24] and PTC [16] respectively. Details about

Table 2. Average classification accuracy (± standard error, when available). The best performing model for each dataset is highlighted in bold.

	D&D	MUTAG	PROTEINS	NCI1	ENZYMES	PTC
GK	75.90 ± 0.10	81.70 ± 2.00	71.70 ± 0.60	-	24.90 ± 0.20	57.30 ± 1.40
WL	79.78 ± 0.36	82.05 ± 0.36	-	82.19 ± 0.18	52.22 ± 1.26	-
DGK	-	87.44 ± 2.72	75.68 ± 0.54	80.31 ± 0.46	53.43 ± 0.91	60.08 ± 2.55
DGCNN	79.37 ± 0.94	85.83 ± 1.66	75.54 ± 0.94	74.44 ± 0.47	-	58.59 ± 2.47
GIN	-	89.00 ± 6.00	75.90 ± 3.80	**82.70 ± 1.60**	-	66.60 ± 6.90
SAGPool	76.45 ± 0.97	-	71.86 ± 0.97	74.18 ± 1.20	-	-
DIFFPool	80.64	-	76.25	-	62.53	-
ABDP	-	91.10 ± 6.70	78.50 ± 2.90	-	64.00 ± 4.00	67.40 ± 4.30
ASAP	76.87 ± 0.70	-	74.19 ± 0.79	71.48 ± 0.42	-	-
VIPool	82.68 ± 4.10	-	**79.91 ± 4.10**	-	57.50 ± 6.10	-
MinCutPool	80.80 ± 2.30	79.90 ± 2.10	76.50 ± 2.60	-	-	-
CGIPool	-	80.65 ± 0.79	74.10 ± 2.31	78.62 ± 1.04	-	-
C2N-ABDP	**86.83 ± 1.15**	**91.50 ± 0.73**	79.00 ± 1.49	80.77 ± 0.85	**64.16 ± 3.85**	**71.06 ± 1.48**

the datasets are shown in Table 1. For the D&D and NCI1 datasets, each pooling layer will reduce the total number of nodes in the graph to 10% of the original. For other datasets, the reduction ratio is 25%. We do 10-fold cross-validation on each dataset and repeat this 10 times. The average accuracy and variance of cross-validation results are shown in Table 2. For all datasets, the number of epochs is selected according to the best cross-validation accuracy.

4.1   Results and Analysis

The experimental results in Table 2 show the advantages of C2N-ADBP, which can be summarized as follows. On all experimental datasets, C2N-ABDP outperforms DIFFPool and ABDP, suggesting the efficacy of the proposed improvement on DIFFPool. While DIFFPool only aggregates nodes based on the probability that the node belongs to the cluster when generating a cluster representation, we use the C2N attention to decouple the node weights from the assignment matrix and distribute the node weights in a more intuitive way. The attention mechanism of ABDP only calculates the weight of nodes based on the attention between nodes, while we propose an attention mechanism that measures the relationship between nodes and the overall information of the cluster where the nodes are located, making it more suitable for graph pooling. We use SVD to extract cluster information in a way that is robust to node ordering, and the experimental results prove that the extracted cluster information is effective.

On both the D&D and PTC datasets, C2N-ABDP achieves state-of-the-art performance, surpassing other competitive models in terms of classification accuracy. This is due to the advantages of the pooling method based on the cluster assignment matrix. In C2N-ABDP, the cluster assignment matrix is used to generate the feature matrix and the adjacency matrix of the next layer, which ensures that C2N-ABDP fully retains the graph structure information while reducing the size of the graph data. However, other models such as SAGPool directly delete nodes, resulting in the loss of graph structure information related to the deleted nodes. In addition, we use a hard assignment matrix to assign nodes to clusters, so that each node can only be assigned to one cluster, which in turn reduces overfitting and makes it possible to use the C2N attention mechanism.

5 Conclusion

In this paper we proposed C2N-ABDP, a pooling method suitable for hierarchical graph pooling architectures. C2N-ABDP is based on DIFFPool but adds an attention mechanism that is able to effectively utilize the graph structure information and distinguish the node importance from the probability that the node belongs to a cluster. To this end, we proposed C2N attention, which enabled our model to better capture the membership of each node in a cluster. C2N-ABDP also uses SVD to extract cluster information that is robust to the order of nodes. The experimental results show that C2N-ABDP can achieve state-of-the-art performance on multiple graph classification datasets.

Acknowledgments. This work is supported by the National Natural Science Foundation of China under Grants T2122020, 61976235, and 61602535. This work is also partly supported by the Program for Innovation Research in the Central University of Finance and Economics.

References

1. Agarap, A.F.: Deep learning using rectified linear units (relu). arXiv preprint arXiv:1803.08375 (2018)
2. Bai, L., Cui, L., Jiao, Y., Rossi, L., Hancock, E.R.: Learning backtrackless aligned-spatial graph convolutional networks for graph classification. IEEE Trans. Pattern Anal. Mach. Intell. **44**(2), 783–798 (2022)
3. Bai, L., et al.: Learning graph convolutional networks based on quantum vertex information propagation (extended abstract). In: 38th IEEE International Conference on Data Engineering, ICDE 2022, Kuala Lumpur, Malaysia, May 9–12, 2022. pp. 3132–3133. IEEE (2022)
4. Bai, L., et al.: Learning graph convolutional networks based on quantum vertex information propagation. IEEE Trans. Knowl. Data Eng. **35**(2), 1747–1760 (2023)
5. Bianchi, F.M., Grattarola, D., Alippi, C.: Spectral clustering with graph neural networks for graph pooling. In: International Conference on Machine Learning, pp. 874–883. PMLR (2020)
6. Bicciato, A., Cosmo, L., Minello, G., Rossi, L., Torsello, A.: Classifying me softly: A novel graph neural network based on features soft-alignment. In: S+SSPR. pp. 43–53. Springer (2022). https://doi.org/10.1007/978-3-031-23028-8_5
7. Borgwardt, K.M., Ong, C.S., Schönauer, S., Vishwanathan, S., Smola, A.J., Kriegel, H.P.: Protein function prediction via graph kernels. Bioinformatics **21**(suppl_1), i47–i56 (2005)

8. Cosmo, L., Minello, G., Bronstein, M., Rodolà, E., Rossi, L., Torsello, A.: Graph kernel neural networks. arXiv preprint arXiv:2112.07436 (2021)
9. Cui, L., Bai, L., Bai, X., Wang, Y., Hancock, E.R.: Learning aligned vertex convolutional networks for graph classification. IEEE Trans. Neural Netw. Learn. Syst. Press, 1808–1822 (2021). https://doi.org/10.1109/TNNLS.2021.3129649
10. Debnath, A.K., Lopez de Compadre, R.L., Debnath, G., Shusterman, A.J., Hansch, C.: Structure-activity relationship of mutagenic aromatic and heteroaromatic nitro compounds. correlation with molecular orbital energies and hydrophobicity. J. Med. Chem. **34**(2), 786–797 (1991)
11. Dobson, P.D., Doig, A.J.: Distinguishing enzyme structures from non-enzymes without alignments. J. Mol. Biol. **330**(4), 771–783 (2003)
12. Gao, H., Ji, S.: Graph u-nets. In: International Conference on Machine Learning, pp. 2083–2092. PMLR (2019)
13. Hamilton, W., Ying, Z., Leskovec, J.: Inductive representation learning on large graphs. In: Advances in Neural Information Processing Systems, vol. 30 (2017)
14. He, K., Zhang, X., Ren, S., Sun, J.: Deep residual learning for image recognition. In: Proceedings of the IEEE Conference on Computer Vision And Pattern Recognition, pp. 770–778 (2016)
15. Kipf, T.N., Welling, M.: Semi-supervised classification with graph convolutional networks. In: 6th International Conference on Learning Representations (2017)
16. Kriege, N., Mutzel, P.: Subgraph matching kernels for attributed graphs. In: Proceedings of the 29th International Conference on International Conference on Machine Learning, pp. 291–298 (2012)
17. Krizhevsky, A., Sutskever, I., Hinton, G.E.: Imagenet classification with deep convolutional neural networks. Commun. ACM **60**(6), 84–90 (2017)
18. LeCun, Y., Kavukcuoglu, K., Farabet, C.: Convolutional networks and applications in vision. In: Proceedings of 2010 IEEE International Symposium on Circuits and Systems, pp. 253–256. IEEE (2010)
19. Lee, J., Lee, I., Kang, J.: Self-attention graph pooling. In: International Conference On Machine Learning, pp. 3734–3743. PMLR (2019)
20. Li, M., Chen, S., Zhang, Y., Tsang, I.: Graph cross networks with vertex infomax pooling. Adv. Neural. Inf. Process. Syst. **33**, 14093–14105 (2020)
21. Liu, Y., Cui, L., Wang, Y., Bai, L.: Abdpool: Attention-based differentiable pooling. In: 2022 26th International Conference on Pattern Recognition (ICPR), pp. 3021–3026. IEEE (2022)
22. Pang, Y., Zhao, Y., Li, D.: Graph pooling via coarsened graph infomax. In: Proceedings of the 44th International ACM SIGIR Conference on Research and Development in Information Retrieval, pp. 2177–2181 (2021)
23. Ranjan, E., Sanyal, S., Talukdar, P.: Asap: Adaptive structure aware pooling for learning hierarchical graph representations. In: Proceedings of the AAAI Conference on Artificial Intelligence. vol. 34, pp. 5470–5477 (2020)
24. Schomburg, I., et al.: Brenda, the enzyme database: updates and major new developments. Nucleic Acids Res. **32**(suppl_1), D431–D433 (2004)
25. Shervashidze, N., Schweitzer, P., Van Leeuwen, E.J., Mehlhorn, K., Borgwardt, K.M.: Weisfeiler-lehman graph kernels. J. Mach. Learn. Res. **12**(9) (2011)
26. Shervashidze, N., Vishwanathan, S., Petri, T., Mehlhorn, K., Borgwardt, K.: Efficient graphlet kernels for large graph comparison. In: Artificial intelligence and statistics, pp. 488–495. PMLR (2009)
27. Vaswani, A.,et al.: Attention is all you need. In: Advances in Neural Information Processing Systems, vol. 30 (2017)

28. Veličković, P., Cucurull, G., Casanova, A., Romero, A., Lio, P., Bengio, Y.: Graph attention networks. In: 6th International Conference on Learning Representations (2017)
29. Wale, N., Watson, I.A., Karypis, G.: Comparison of descriptor spaces for chemical compound retrieval and classification. Knowl. Inf. Syst. **14**, 347–375 (2008)
30. Xu, K., Hu, W., Leskovec, J., Jegelka, S.: How powerful are graph neural networks? In: 7th International Conference on Learning Representations, ICLR 2019, New Orleans, LA, USA, May 6–9, 2019 (2019)
31. Yanardag, P., Vishwanathan, S.: Deep graph kernels. In: Proceedings of the 21th ACM SIGKDD International Conference on Knowledge Discovery and Data Mining, pp. 1365–1374 (2015)
32. Ying, Z., You, J., Morris, C., Ren, X., Hamilton, W., Leskovec, J.: Hierarchical graph representation learning with differentiable pooling. In: Advances in Neural Information Processing Systems, vol. 31 (2018)
33. Zhang, M., Cui, Z., Neumann, M., Chen, Y.: An end-to-end deep learning architecture for graph classification. In: Proceedings of the AAAI Conference On Artificial Intelligence, vol. 32 (2018)
34. Zhang, Z., et al.: Hierarchical graph pooling with structure learning. arXiv preprint arXiv:1911.05954 (2019)

Splitting Structural and Semantic Knowledge in Graph Autoencoders for Graph Regression

Sarah Fadlallah[iD], Natália Segura Alabart[iD], Carme Julià[iD], and Francesc Serratosa[✉][iD]

Research Group ASCLEPIUS: Smart Technology for Smart Healthcare, Department D'Enginyeria Informática I Matemátiques, Universitat Rovira I Virgili, 43k007 Tarragona, Catalonia, Spain
{sarah.fadlallah,natalia.segura,carme.julia,francesc.serratosa}@urv.cat

Abstract. This paper introduces ReGenGraph, a new method for graph regression that combines two well-known modules: an autoencoder and a graph autoencoder. The main objective of our proposal is to split the knowledge in the graph nodes into semantic and structural knowledge during the embedding process. It uses the autoencoder to extract the semantic knowledge and the graph autoencoder to extract the structural knowledge. The resulting embedded vectors of both modules are then combined and used for graph regression to predict a global property of the graph. The method demonstrates improved performance compared to classical methods, i.e., autoencoders or graph autoencoders alone. The approach has been applied to predict the binding energy of chemical compounds represented as attributed graphs but could be used in other fields as well.

Keywords: Graph embedding · Graph Convolutional Networks · Autoencoders · Graph Autoencoders · Graph Regression

1 Introduction

A graph, in general, is a data structure depicting a collection of entities represented as nodes, and their pairwise relationships represented as edges. There is a growing interest in having graph-based techniques applied to machine learning, for instance, in biotechnology, they are used to represent chemical compounds in order to predict their toxicity [2]. This can be attributed to their effectiveness in characterising instances of data with complex structures and rich attributes. An example of this is the ability of the Graph Edit Distance to capture the dissimilarity between graphs [15,16].

In this paper, we propose a computational method called ReGenGraph: Regression on Generated Graphs. This method applies graph regression techniques based on graph autoencoders (GAE) (e.g., [6,8]) to predict a global property of the graph. The key point of ReGenGraph is to split the knowledge in a

M. Vento et al. (Eds.): GbRPR 2023, LNCS 14121, pp. 81–91, 2023.
https://doi.org/10.1007/978-3-031-42795-4_8

graph into semantic and structural knowledge during the embedding process. This means that the global property is predicted in two steps. First, an autoencoder and a GAE are used to deduce a latent vector. Then, the global property of the graph is computed through a regression module applied to the resulting latent vector.

As a result of ReGenGraph's ability to split node attributes into two types, we saw that our developed model would be suitable for data where both structural and semantic knowledge are present, such as the case with molecular graphs. They make a good example of a quite natural way to describe a set of atoms and their interactions [2, 13]. In this case, we use the ReGenGraph to predict the binding energy of chemical compounds depicted as graphs that represent their atoms and bonds as nodes and edges respectively.

In the next section, we present a summarised overview of the techniques involved in this work. In Sect. 3 we explain our approach in detail. In Sect. 4 we show the experimental validation we carried out, concluding the paper with Sect. 5.

2   Related Work

2.1   Autoencoders

Autoencoders are a particular class of neural networks that are employed in machine learning to capture the most basic representations of an entity. To achieve this, they are trained to reconstruct the input data after having generated an intermediate data called latent space [7]. Autoencoders can be used for dimensionality reduction, data denoising, or anomaly detection [3]. The obtained intermediate representations can also be used as learning tokens for classification and prediction tasks, or for the generation of synthetic data.

An autoencoder consists of two components: an encoder that converts the input space into a latent space, Z, and a decoder that converts the lower-dimensional representation back to the original input space. W_0, and W_1 can be defined as the trainable weights of the encoder and decoder, respectively. The encoding processes results in latent vector z for each entity n in a latent space, commonly denoted as $Z \in \mathbb{R}^{n \times a}$, defined by the number of entities n, e.g., atoms in a molecule, and the features extracted in the latent space, a. Both encoders and decoders include non-linear activation functions. This non-linearity typically increases the expressive ability of the network and enables it to learn a range of tasks at various levels of complexity.

From there, autoencoders continued to evolve and various enhancements were introduced to tackle different issues and improve their representative capabilities. An example of this is the denoising autoencoder, a variation that adds noise to input, corrupting some of the samples on purpose, in order to prevent the network from learning the identity function, i.e., having the network learn the data points themselves rather than their representation [17]. Another popular architecture was introduced by Diederik P. Kingma and Max Welling known as the variational autoencoder [4]. They combine autoencoders with probabilistic Bayesian inference to map inputs according to their distribution in the latent space.

2.2   Graph Autoencoders

There has been a growth in using neural networks on data represented as graphs across various domains, despite the complexity of graphs that results from their intertwined characteristics. For the scope of this work, we focus on applications concerning drug potency prediction [1]. The currently used techniques can be divided into four categories, recurrent graph neural networks, convolutional graph neural networks, graph autoencoders, and spatial-temporal graph neural networks [19]. Graph Convolutional Networks (GCNs) can be an especially attractive choice since they can capture and leverage the structure of a given graph, making them well-suited for tasks where the topology of the graph is crucial such as is the case with the analysis and generation of chemical compounds.

A graph with attributes, represented by a node attribute matrix X, and an adjacency matrix A, can be represented as $\mathbf{G(X,A)}$, where $X \in \mathbb{R}^{n \times f}$ is a matrix of size $n \times f$, with n being the number of nodes and f being the number of attributes or node features. The adjacency matrix $A \in \mathbb{R}^{n \times n}$ is of size $n \times n$, where the $A_{i,j} = 1$ if there is an edge between the ith and the jth node and 0 otherwise. The graph's edges are unattributed and undirected, meaning that if there is an edge from node i to node j, there is also an edge from node j to node i, which is represented by the equality $A_{i,j} = A_{j,i}$.

GAEs are based on the concept of a GCN, which in turn is built on the notion of generalising convolution-like processes on normal grids, $e.g.$, images, to graph-structured data through neural network layers [6].

The key idea behind GCNs is to define the neighborhood of a node in the graph, using the information from the neighboring nodes to update the node's representation. This can be accomplished by defining a convolution operation on the graph, which is typically implemented as a weighted sum of the representations of the neighboring nodes. A learnable weight matrix is often used to determine the weights of this sum, which the network learns as it updates the node's representation. Node attributes can also be used to infer global properties about the graph's structure and the links between its nodes.

Just like the classical autoencoder, GAEs are composed of two main parts: an encoder and a decoder. The encoder embeds input graphs through a GCN as defined in [6] returning a latent matrix $Z \in \mathbb{R}^{n \times b}$ with the graph's unique properties. The number of features in the latent space is b. Equation 1 shows the encoder's function:

$$Z = GCN(X, A) = \tilde{A}ReLU\left(\tilde{A}XW_0'\right)W_1' \tag{1}$$

where $\tilde{A}$ is a symmetrically normalised adjacency matrix computed from A, while W_0' and W_1' are the weight matrices for each layer, which are learned through a learning algorithm. Note that $ReLU$ is the classical non-negative linear equation.

The decoder is defined as Eq. 2:

$$A^* = \sigma\left(ZZ^T\right) \tag{2}$$

where $\sigma\left(\cdot\right)$ is the sigmoid function, T means the transposed matrix. The output, A^*, is a matrix of real numbers between 0 and 1 that represents the probability of an existing edge in the reconstructed adjacency matrix. Note that in order to deduce the final reconstructed matrix, a round function is applied to A^* to discern between non-edge and edge, i.e., zero and one values.

As the aim of the GAE is to reconstruct the adjacency matrix such that it is similar to the original one, the learning algorithm minimises the mean square distance between these matrices defined by Eq. 3,

$$\mathcal{L} = \frac{1}{n^2}\sum_{i=1}^{n}\sum_{j=1}^{n} w_{pos}A_{i,j}logA^*_{i,j} + w_{neg}(1 - A_{i,j})log(1 - A^*_{i,j}) \qquad (3)$$

where w_{pos} and w_{neg} are introduced to deal with the value imbalance between pairs of nodes with an edge and pairs of nodes without an edge.

3   Proposed Approach: ReGenGraph

The basis of the GAE approaches is the constraint that knowledge associated with nodes is related to knowledge attached to edges and vice versa [6]. That is, it is assumed that there is a relation between the local structural pattern and the node attributes.

We have designed a specific model based on GAE that handles graphs with nodes that have these two types of attributes: those that are impacted by structural patterns and those that are not related to edges. Specifically, our approach is based on two modules that work accordingly. The first one is an autoencoder [9] that captures semantic information, without structural relations but rather by only utilising certain node attributes. The other module is a GAE [6] that captures structural knowledge, which is achieved by exploiting the remaining node attributes and edges. Both modules project their data into a latent domain, which is then used for any fitting mechanism, regression was chosen for this case. The ReGenGraph architecutre is depicted in Fig. 1. We use the GAE defined in [5] and summarised in Sect. 2.2. It is important to note that both the autoencoder and the GAE are used for extracting features during the encoding stage in order to be used for a prediction or classification task. Nevertheless, the complete models, with both encoder and decoder components, are also useful for reconstructing the graph.

The decision on which node attributes to use in the autoencoder and which to use in the GAE is made through a validation process. This could mean training the model on different subsets of selected attributes, i.e., randomly selecting attributes for each architecture and determining the combination that results in the lowest loss for both. However, for specific tasks, one can make this decision based on their field knowledge of the problem. For the sake of this experiment, as a consequence of the low number of features we are working with (only the atom coordinates, and its atomic number), it seemed intuitive to cast the cartesian coordinates to the GAE as structural features, while feeding the atomic number to the autoencoder as node semantic feature.

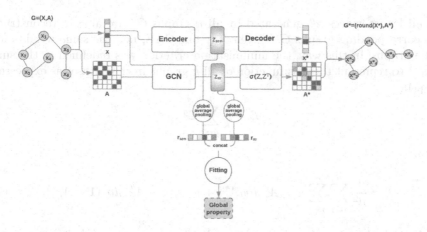

Fig. 1. Schematic view of our architecture for graph regression based on an autoencoder, a graph autoencoder, fitted with a regression module.

The latent space of our architecture is created by combining the latent space of the autoencoder, represented as Z_{sem}, and the latent space of the GAE, represented as Z_{str}. Graphs are structures that must be invariant to the order of the nodes, meaning they have the property of being node-position invariant. A common way to achieve this property is by computing the sum, mean, minimum, or maximum of each feature for all nodes. We have chosen to calculate the mean, as it makes the architecture independent of the number of nodes. Applying this mean is commonly known as the global average pooling. Then, given z_{str} vectors, the r_{str} vector is generated by computing their mean. Note the length of the vector r_{str} is independent of the number of nodes n. This is an important feature since it means that we can fit the system with graphs that have different numbers of nodes.

Finally, the concatenated vector composed of r_{sem} and r_{str} is used for regression fitting. This vector is used to determine the global property of the graph.

3.1 The Learning Process

The learning process is achieved in two steps. Initially, both weights W_0, W_1 of the autoencoder, and W_0', W_1' in the GAE are learned given all graphs G^g, where $g = 1, ..., k$. Following that, the regression weights are learned, given the returned latent vectors z_{sem}^g and z_{str}^g of all graphs G^g in the training set, where $g = 1, ..., k$. For the scope of this paper and its application, we focus on GAEs. More on the learning process of the autoencoder, weights W_0, and W_1, can be found in the original work [7].

GAEs (Sect. 2.2) were modeled to reconstruct only one, usually huge, graph. Thus, the aim of the learning process, which minimises Eq. 3, is to reconstruct this unique graph. In that case, z would have to be defined such that it resembles the inherent properties of this graph. We are in a different scenario. We wish that all latent spaces z^g generated by all k graphs G^g are able to reconstruct their corresponding graphs G^{*g} given only one GAE, i.e., the same weights for all the graphs. In this way, the minimisation criterion was redefined as the sum of Eq. 3 to represent the loss function of all k graphs in the dataset as expressed in Eq. 4:

$$\mathcal{L} = \frac{1}{k} \sum_{g=1}^{k} \mathcal{L}^g \qquad (4)$$

where,

$$\mathcal{L}^g = \frac{1}{n^2} \sum_{i=1}^{n} \sum_{j=1}^{n} w_{pos} A_{i,j}^g log A_{i,j}^{*g} + w_{neg}(1 - A_{i,j}^g)log(1 - A_{i,j}^{*g}). \qquad (5)$$

describes the loss function per each graph G^g, where w_{pos} and w_{neg} represent the positive and negative weights.

4   Experimental Validation

ReGenGraphs have been applied to predict the interaction or binding energy between a ligand and its corresponding protein. This binding energy is essentially the free energy of the complex AB or the Gibbs energy, which is calculated by subtracting the Gibbs energy of molecule A, i.e., the ligand, and the Gibbs energy of molecule B, i.e., the protein, from the Gibbs energy of the complex AB.

As mentioned above, our proposal is thought for data that presents both semantic and structural knowledge. In the database used in the experiments, the node attributes consist of the three-dimensional position of the atom and its atomic number. In the case of the first attribute, we can clearly see a relation between having a bond, an edge in the graph, between two atoms, two nodes in the graph, and the proximity between these atoms. Contrarily, in the case of the second attribute, there is no relation between the type of atom and being connected to a similar one. The contrary option would be, for instance, that oxygen tends to be connected to oxygen and not to other atoms.

4.1   Database

We utilized data from the Immune Epitope Database (IEDB) [18] to develop a new database. IEDB database is a valuable resource for studying specific diseases, as it offers researchers the ability to identify and analyze epitopes that are relevant to their particular research goals [10] or for training and developing web servers aimed at predicting binding interactions between peptides and major histocompatibility complex molecules [12].

This new database selectively includes the HLA-A02:01 allele and peptides of length 9, known as nonamers. The choice of HLA-A02:01 was made as it is one of the most prevalent and polymorphic subtypes of HLA-A molecules both in humans and in the IEDB. The choice of nonamers was made as it is the most frequent peptide length that binds to HLA-A*02:01 allele in the IEDB database. A total of 4872 peptides were selected and then further filtered to yield a total of 500 graphs. This dataset was created using the data published up until 23/10/2022.

The database was employed to generate 3D compounds utilizing FoldX [14], a software that predicts the stability of protein structures and mutations and gives their binding energies. The 3D compounds were created with the functions *RepairPDB*, *BuildModel* and *AnalyseComplex* from FoldX. The 3D compounds provide a more detailed insight into the interactions between the peptides and the HLA-A02:01 allele. The 5ENW structure of HLA-A02:01 served as the template to create all 3D compounds [11]. The 3D compounds are complexes involving an HLA molecule and a peptide of interest. In this paper, we solely utilized the structure and coordinates of the peptide to reduce the complexity associated with studying the entire complex.

4.2   Architecture Configuration

The autoencoder was modeled with a fully connected neural network, which only has one hidden layer with 20 neurons that takes the adjacency matrix and atomic number as input. The length of z_{sem} is 20. The input and output layers have 96 neurons to consider the largest graph in the set (graphs varied between 48 and 96 nodes per graph with an average of 71 nodes). The weights were defined as follows, $W_0 \in \mathbb{R}^{96 \times 20}$ and $W_1 \in \mathbb{R}^{20 \times 96}$. The hidden layer applied a sigmoid activation function while the output layer utilised a linear function. The back-propagation algorithm was used for learning.

The input X of the GAE is composed of a matrix of 96 (number of nodes) times 3 (3D position). The input A of the GAE is composed of a square matrix of 96 times 96. $W_0' \in \mathbb{R}^{3 \times 100}$ and $W_1' \in \mathbb{R}^{100 \times 20}$ took their shape corresponding to the size of the hidden layer (100), and the number of attributes given (3). z_{sem} is a matrix of 96 times 20 and thus r_{str} is a vector that has a length of 20. For graphs with fewer than 96 nodes, we fill the remaining rows of X and the rows and columns of A with zeros. This procedure is applied to both the autoencoder and the GAE.

Finally, the fitting function is modeled by a classical regression. Thus, it receives a vector of 40 elements, composed by concatenating the 20 elements from r_{sem} and their counterparts from r_{str}. The function outputs only one real number representing the binding energy.

4.3   Binding Energy Prediction

We want to heuristically validate the need of using an autoencoder and a GAE, instead of applying a classical scheme that is composed of only one of them, namely, an autoencoder or GAE.

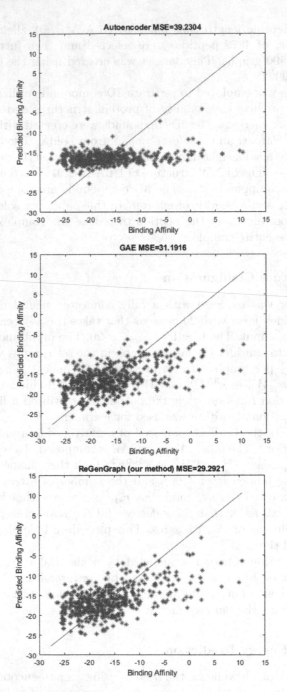

Fig. 2. Three scatter plots showing the predicted and experimental binding energy of the compounds in the database. From top to down, using only an Autoencoder, using only a GAE and our model. The mean square errors appear on the top of the scatters.

Figure 2 shows three scatter plots of computed and experimental binding energy values corresponding to the compounds in the database. In the first case, only an autoencoder and a fitting function were used. Note that in this scenario, it was not possible to reconstruct the bonds of the compounds. In the second case, only a GAE and a fitting function were used. In this scheme, the compound can be reconstructed. Finally, our method was applied by combining latent representations derived from both the autoencoder and the GAE to be used by the fitting function.

The first technique returns the highest mean square error (MSE), with a value of 39.23. Following that, the GAE technique demonstrates an MSE of 31.19, and our proposed method achieves an MSE of 29.29. These outcomes validate the proposed method in a heuristic manner, which is based on splitting the node attributes -features of the atom- in two parts, one that is independent of the graph edges -existence and type of a bond-N, while the other remained dependent on them. Hence, by carefully deciding which attributes to be taken into account and which ones to be discarded, we have demonstrated that it is worthwhile to define a dual model that applies this split of attributes.

5 Conclusions

The binding energy between a protein and its ligand is key to understanding the dynamics and stability of the protein-ligand complex. The non-covalent interactions between the two molecules are crucial for predicting the strength of the protein-ligand complex and identifying drug candidates that selectively bind to the target proteins in the drug discovery process.

Given that we currently have some experimental data, this aim can be achieved by modern computational methods based on machine learning. A binding energy predictor of drugs has been presented, which is based on two steps. The first part is the conversion of the molecule into the interaction graph. The second is a new architecture composed of an autoencoder, a graph autoencoder, and a regression module.

A key aspect of our approach is the separation of the semantic and the structural knowledge of the compounds. The first is processed through the autoencoder while the second is processed through the graph autoencoder. This main feature is independent of the application, which means, our proposal could have different applications in other fields. The only important aspect to be considered is discerning between attributes that are dependent on the structure and attributes that are not.

Practical experiments show the ability of our method to predict the binding energy. In addition to that, they also show that the mean square error of the binding energy prediction using a graph autoencoder is larger than using our method.

In future work, we plan to test our proposal by using different architectures for the autoencoder in addition to applying other fitting functions in the regression model, such as neural networks. we also aim to further validate our method by testing it on other graphs and datasets. Despite the simplicity of the chosen functions, the results are promising.

Acknowledgements. This research is supported by the Universitat Rovira i Virgili through the Martí Franquès grant and partially funded by AGAUR research group 2021SGR-00111: "ASCLEPIUS: Smart Technology for Smart Healthcare".

References

1. Fadlallah, S., Julià, C., Serratosa, F.: Graph regression based on graph autoencoders. In: Krzyzak, A., Suen, C.Y., Torsello, A., Nobile, N. (eds.) Structural, Syntactic, and Statistical Pattern Recognition, pp. 142–151. Springer International Publishing, Cham (2022)
2. Garcia-Hernandez, C., Fernández, A., Serratosa, F.: Ligand-based virtual screening using graph edit distance as molecular similarity measure. J. Chem. Inf. Model. **59**(4), 1410–1421 (2019)
3. Goodfellow, I., Bengio, Y., Courville, A.: Deep Learning. MIT Press (2016). http://www.deeplearningbook.org
4. Kingma, D.P., Rezende, D.J., Mohamed, S., Welling, M.: Semi-supervised learning with deep generative models (2014)
5. Kipf, T.N.: Deep Learning with Graph-Structured Representations. Ph.D. thesis, University of Amsterdam (2020)
6. Kipf, T.N., Welling, M.: Semi-supervised classification with graph convolutional networks. CoRR abs/1609.02907 (2016). arXiv:1609.02907
7. Kramer, M.A.: Nonlinear principal component analysis using autoassociative neural networks. AIChE J. **37**, 233–243 (1991)
8. Le, T., Le, N., Le, B.: Knowledge graph embedding by relational rotation and complex convolution for link prediction. Expert Syst. Appl. **214**, 119122 (2023). https://doi.org/10.1016/j.eswa.2022.119122
9. Majumdar, A.: Graph structured autoencoder. Neural Netw. **106**, 271–280 (2018). https://doi.org/10.1016/j.neunet.2018.07.016
10. Naveed, M., et al.: A reverse vaccinology approach to design an mrna-based vaccine to provoke a robust immune response against hiv-1. Acta Biochimica Polonica **70**(2) (2023). https://doi.org/10.18388/abp.2020_6696
11. Remesh, S., et al.:Unconventional peptide presentation by major histocompatibility complex (mhc) class i allele hla-a*02:01:breaking confinement. J. Biol. Chem. **292**(13) (2017). https://doi.org/10.1074/jbc.M117.776542
12. Reynisson, B., Alvarez, B., Paul, S., Peters, B., Nielsen, M.: Netmhcpan-4.1 and netmhciipan-4.0: improved predictions of MHC antigen presentation by concurrent motif deconvolution and integration of MS MHC eluted ligand data. Nucleic Acids Res. **48**(W1), W449–W454 (2020). https://doi.org/10.1093/nar/gkaa379
13. Rica, E., Álvarez, S., Serratosa, F.: Ligand-based virtual screening based on the graph edit distance. Int. J. Mol. Sci. **22**(23), 12751 (2021)
14. Schymkowitz, J., Borg, J., Stricher, F., Nys, R., Rousseau, F., Serrano, L.: The foldx web server: an online force field. Nucleic acids research 33(Web Server issue) (2005). https://doi.org/10.1093/nar/gki387
15. Serratosa, Francesc: Redefining the graph edit distance. SN Comput. Sci. **2**(6), 1–7 (2021). https://doi.org/10.1007/s42979-021-00792-5

16. Serratosa, F., Cortés, X.: Graph edit distance: moving from global to local structure to solve the graph-matching problem. Pattern Recogn. Lett. **65**, 204–210 (2015)
17. Vincent, P., Larochelle, H., Lajoie, I., Bengio, Y., Manzagol, P.A.: Stacked denoising autoencoders: Learning useful representations in a deep network with a local denoising criterion. J. Mach. Learn. Res. **11**, 3371–3408 (2010)
18. Vita, R., et al.: The immune epitope database (iedb): 2018 update. Nucleic Acids Res. **47**(D1) (2018). https://doi.org/10.1093/nar/gky1006
19. Wu, Z., Pan, S., Chen, F., Long, G., Zhang, C., Yu, P.S.: A comprehensive survey on graph neural networks. IEEE Trans. Neural Netw. Learn. Syst. **32**(1), 4–24 (2021). https://doi.org/10.1109/TNNLS.2020.2978386

Graph Normalizing Flows to Pre-image Free Machine Learning for Regression

Clément Glédel[(✉)], Benoît Gaüzère, and Paul Honeine

Univ Rouen Normandie, INSA Rouen Normandie, Université Le Havre Normandie,
Normandie Univ, LITIS UR 4108, 76000 Rouen, France
`clement.gledel@univ-rouen.fr`

Abstract. In Machine Learning, data embedding is a fundamental aspect of creating nonlinear models. However, they often lack interpretability due to the limited access to the embedding space, also called latent space. As a result, it is highly desirable to represent, in the input space, elements from the embedding space. Nevertheless, obtaining the inverse embedding is a challenging task, and it involves solving the hard pre-image problem. This task becomes even more challenging when dealing with structured data like graphs, which are complex and discrete by nature. This article presents a novel approach for graph regression using Normalizing Flows (NFs), in order to avoid the pre-image problem. By creating a latent representation space using a NF, the method overcomes the difficulty of finding an inverse transformation. The approach aims at supervising the space generation process in order to create a space suitable for the specific regression task. Furthermore, any result obtained in the generated space can be translated into the input space through the application of the inverse transformation learned by the model. The effectiveness of our approach is demonstrated by using a NF model on different regression problems. We validate the ability of the method to efficiently handle both the pre-image generation and the regression task.

Keywords: Graph Normalizing Flows · Pre-image problem · Regression · Interpretability · Nonlinear embedding

1 Introduction

Graph machine learning generally operates by embedding graph data to a meaningful space known as the latent (or feature) space. This embedding can either be implicit, as in the case of kernel machines, or explicit through the use of nonlinear operations in deep neural networks or more classic graph embedding approaches [3,11]. While providing accurate prediction models for classification or regression tasks, such methods lack interpretability, and it may be interesting to invert the embedding and map the results back to the graph space to analyze the behavior of the model. This process is referred to as the pre-image problem. Several solutions for the pre-image problem have been proposed in the general

M. Vento et al. (Eds.): GbRPR 2023, LNCS 14121, pp. 92–101, 2023.
https://doi.org/10.1007/978-3-031-42795-4_9

case [1,9]. Moreover, some solutions have been proposed to solve the pre-image problem on graph data [2,10].

In this work, we propose to design an interpretable prediction model using a graph regression method that addresses the issue of the pre-image. The key idea is to define a nonlinear-embedding function that is invertible by design. The learned embedding space is designed to linearly organize the samples, leading to good regression performances by the application of any standard linear regression method in this space. Additionally, pre-images can be conveniently generated by applying the inverse mapping on any sample of interest from the latent space.

Our approach producing a reversible nonlinear-embedding function takes inspiration from recent advances in graph generative models [5–7], including Normalizing Flows (NFs) for graphs and molecular data [20]. By defining an invertible transformation from the complex distribution of input data to a simple distribution easy to manipulate, NFs can learn a latent space while guaranteeing the invertibility of the model. Our experimental results showcase the efficacy of our proposed methodology through the use of the MoFlow architecture [20], a graph normalizing flow (GraphNF) using coupling-layers to operate on graphs represented by a combination of a feature matrix and adjacency tensor. We conducted experiments on well-known molecular datasets to demonstrate the applicability of our approach in addressing graph regression tasks while producing a high-quality representation space free of the pre-image problem.

The paper is organized as follows: Sect. 2 gives an overview of NFs and Graph-NFs. Our contributions are presented in Sect. 3, which is divided into three parts. We first revisit the NF to address a regression task, and then detail the regression model. Lastly, we introduce the pre-image generation operations. The experiments and conclusion follow in Sects. 4 and 5, respectively.

2   Normalizing Flow Preliminaries

Normalizing Flows (NFs) are generative models that learn an invertible transformation function Φ between two probability distributions: a complex data distribution $P_{\mathcal{X}}$ and another distribution $P_{\mathcal{Z}}$, often chosen as a simple Gaussian distribution represented in a latent space. This allows fast and efficient data generation by sampling from the Gaussian distribution in the latent space and using the inverse function Φ^{-1} to generate data in the input space $\mathcal{X}$. The relationship between the two probability densities in NFs is defined using the change of variable formula. Therefore, considering the input samples $x_1, x_2, \ldots, x_N \in \mathcal{X}$, the training is performed by maximizing the log-likelihood function

$$\log P_{\mathcal{X}}(X) = \sum_{i=1}^{N} \log P_{\mathcal{Z}}(\Phi(\mathbf{x}_i)) + \log \left| \det \left(\frac{\partial \Phi(\mathbf{x}_i)}{\partial \mathbf{x}_i} \right) \right|, \qquad (1)$$

where the determinant of the Jacobian of the function Φ, evaluated at $\mathbf{x}_i$, indicates the degree of deformation between the two distributions. This expression represents the exact relationship between the distributions, which differs from

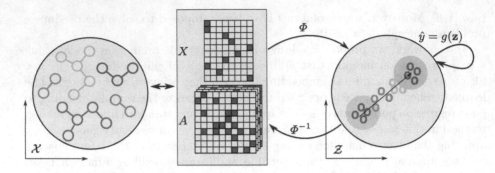

Fig. 1. NF adapted to a regression task, colors correspond to y values.

variational auto-encoders that rely on lower bounds. To define this relationship, the NF model should have an easy-to-invert structure with an easy-to-compute determinant. In order to enhance the model's expressiveness, the transformation Φ combines ℓ bijective functions, i.e., $\Phi = \Phi_\ell \circ \Phi_{\ell-1} \circ \cdots \circ \Phi_1$. This results in computing the Jacobian determinant of Φ as the product of the determinants of all Φ_i. Various NF architectures have been proposed in the literature to satisfy these constraints by defining a triangular Jacobian matrix whose determinant computation is very efficient [14].

Graph Normalizing Flows (GraphNFs) apply the concept of NFs to graph-structured data. While some GraphNFs generate graphs sequentially, such as GraphAF [19] based on a flow-based autoregressive model, a large number of GraphNFs generate graphs in a one-shot manner [8,15,16,20]. In the latter, the first attempt to design a graph neural network using NF structures was GNF [15] where the node features are updated using reversible message passing transformation based on coupling layers. GraphNVP [16] and MoFlow [20] are GraphNFs working in a one-shot manner and representing molecular graphs as a pair of node feature matrix and adjacency tensor. They are based on the use of affine-coupling layers to both the node feature matrix and the adjacency tensor. Specifically, MoFlow involves a modified version of Glow [13] – which is a convolutional NF for image data – to model the bonds and a new graph conditional flow to model the atoms given the bonds using Relational Graph Convolutional Network models. Moreover, to ensure the validity of the generated molecules, MoFlow applies a post-hoc correction step.

3   Proposed Approach

This article presents a graph regression approach based on the NF formalism, where the NF generated latent space $\mathcal{Z}$ is both relevant to the regression task at hand and, by design, does not suffer from the pre-image problem. The underlying idea, illustrated in Fig. 1, involves the three steps. First, we propose to supervise the learning of the NF function $\Phi : \mathcal{G} \rightarrow \mathcal{Z}$ to generate a distribution that follows multiple Gaussian distributions linearly organized in $\mathcal{Z}$. Second, using the learned latent space $\mathcal{Z}$, we can perform straightforward operations (e.g., ridge regression) or more sophisticated algorithms to predict a quantitative value given a data. Finally, as our model is invertible, both the transformation Φ and its inverse Φ^{-1} are produced by our training algorithm. Therefore, it is possible to compute the pre-image of any point from the latent space $\mathcal{Z}$.

3.1 Regression-Based NF

Traditional NF models embed data in such a way that the distribution in the latent space follows a target probability density, typically a Gaussian distribution. However, this configuration does not permit data to be organized based on their quantitative values. In this section, we propose an adaptation of the NF objective function to embed data based on their quantitative values, thus, suitable for linear regression.

Consider a dataset $\mathcal{D} = \{(G_1, y_1), (G_2, y_2) \ldots (G_N, y_N)\}$ composed of input graph data denoted by $G_i \in \mathcal{G}$ and their corresponding quantitative labels denoted by $y_i \in Y \subset \mathbb{R}$. Specifically, we consider the case where every graph is partitioned into its corresponding feature matrix and adjacency tensor, i.e., $G = (\mathbf{X}, \mathbf{A}) \in \mathbb{R}^{n \times d} \times \mathbb{R}^{n \times n \times e}$ where each graph is represented by n nodes of d dimensions and a set of edges characterized by e dimensions. Thus, the total dimension number is $D = n^2 \times e + n \times d$. In this paper, we consider using NF models to represent data, where each data point is represented by a two-part latent representation that corresponds to the features matrix and adjacency tensor. In particular, we concatenate the flattened representations of these two parts to obtain the representation of data in the latent space $\mathcal{Z} \subset \mathbb{R}^D$.

We constitute our latent space using Gaussian distributions, each parameterised by a mean $\boldsymbol{\mu}$ and a covariance matrix $\boldsymbol{\Sigma}$, namely

$$P_{\mathcal{Z}}(\mathbf{z}, \boldsymbol{\mu}, \boldsymbol{\Sigma}) = \frac{1}{\sqrt{(2\pi)^D \det(\boldsymbol{\Sigma})}} e^{-\frac{1}{2}(\mathbf{z} - \boldsymbol{\mu})^\top \boldsymbol{\Sigma}^{-1}(\mathbf{z} - \boldsymbol{\mu})}.$$

From this, we define the log-probability to belong to a Gaussian as

$$\log P_{\mathcal{Z}}(\mathbf{z}, \boldsymbol{\mu}, \boldsymbol{\Sigma}) = -\tfrac{1}{2}\left(D \log(2\pi) + (\mathbf{z} - \boldsymbol{\mu})^\top \boldsymbol{\Sigma}^{-1}(\mathbf{z} - \boldsymbol{\mu})\right) - \log(\det(\boldsymbol{\Sigma})). \quad (2)$$

While common NFs rely on isotropic multivariate Gaussian distributions, thus using a parameterization of zero-valued $\boldsymbol{\mu}$ for all dimensions and the identity matrix as the covariance matrix $\boldsymbol{\Sigma}$, our approach aims at solving regression problems by the use of Gaussian distribution interpolations in $\mathcal{Z}$. The principle is to learn a distribution that spreads along a main axis by interpolating between two Gaussians, which are associated with the extreme quantitative values. Therefore, two Gaussian distributions are defined and parameterized by $(\boldsymbol{\mu}_1, \boldsymbol{\Sigma}_1)$ and $(\boldsymbol{\mu}_2, \boldsymbol{\Sigma}_2)$, which are respectively associated with the minimum and maximum values of Y. For the sake of simplicity, we use isotropic Gaussians, namely with covariance matrices $\boldsymbol{\Sigma}_1 = \boldsymbol{\Sigma}_2 = \sigma^2 \mathbb{I}_D$, where $\mathbb{I}_D$ is the $D \times D$ identity matrix and $\sigma^2 \in \mathbb{R}$ represents the distribution variance.

To carry out the interpolation process, a belonging coefficient τ_{y_i} is assigned to each sample (G_i, y_i) based on its quantitative value computed with

$$\tau_{y_i} = \frac{y_i - \min(Y)}{\max(Y) - \min(Y)}. \tag{3}$$

The interpolated Gaussian mean $\boldsymbol{\mu}_{y_i}$ is computed using

$$\boldsymbol{\mu}_{y_i} = \tau_{y_i}\,\boldsymbol{\mu}_1 + (1 - \tau_{y_i})\,\boldsymbol{\mu}_2. \tag{4}$$

For the interpolation method to be useful, the 2 extreme Gaussian locations in $\mathcal{Z}$ represented by their means $(\boldsymbol{\mu}_1, \boldsymbol{\mu}_2)$ should be distinct and sufficiently separated. Thus, we propose to learn their means within the NF training. To achieve this, we incorporate an additional objective function into the NF objective function, aimed at maximizing the separability of the Gaussians.

Thus, the proposed NF loss function is composed of two terms. The first one applies the change of variable formula (1) to describe the change in density of a single sample. Specifically, for each sample G_i, the corresponding loss is

$$\mathcal{L}_{\text{nf}}(G_i, \boldsymbol{\mu}_{y_i}) = -\log P_{\mathcal{Z}}(\Phi(G_i), \boldsymbol{\mu}_{y_i}, \boldsymbol{\Sigma}_{y_i}) - \log\left|\det\left(\frac{\partial\Phi(G_i)}{\partial G_i}\right)\right|.$$

Here, $\log P_{\mathcal{Z}}(\Phi(G_i), \boldsymbol{\mu}_{y_i}, \boldsymbol{\Sigma}_{y_i})$ refers to (2), which uses the interpolated Gaussian parameters $(\boldsymbol{\mu}_{y_i}, \boldsymbol{\Sigma}_{y_i})$. The mean $\boldsymbol{\mu}_{y_i}$ is computed using (4), while the covariance matrix $\boldsymbol{\Sigma}_{y_i}$ is defined in the same way as the other Gaussian distributions, namely $\boldsymbol{\Sigma}_{y_i} = \sigma^2\,\mathbb{I}_D$. The second term promotes the separation between the two extreme Gaussians with $\mathcal{L}_\mu = -\log\left(1 + \|\boldsymbol{\mu}_1 - \boldsymbol{\mu}_2\|_2^2\right)$. Thus, the final loss function is

$$\mathcal{L}(G_i, y_i) = \mathcal{L}_{\text{nf}}(G_i, \boldsymbol{\mu}_{y_i}) + \beta\mathcal{L}_\mu, \tag{5}$$

with β corresponding to the tradeoff coefficient between the two terms.

Let Θ denotes the set of parameters of Φ, and let ω the set of optimizable parameters, i.e., $\omega = \{\Theta, \boldsymbol{\mu}_1, \boldsymbol{\mu}_2\}$. To estimate the parameters in ω, we employ a stochastic gradient descent algorithm that minimizes the loss function (5) over a randomly selected batch I of the training dataset at each iteration, namely

$$\omega \leftarrow \omega - \eta \sum_{k \in I} \nabla_\omega \mathcal{L}(G_k, y_k), \tag{6}$$

where η is the learning rate.

3.2   Operating in $\mathcal{Z}$

Our approach allows a customized generation of a latent space $\mathcal{Z}$ where data are linearly organized. Therefore, a simple and efficient predictive model can be defined in $\mathcal{Z}$. We denote $g : \mathcal{Z} \to \mathbb{R}$ a linear predictive model. Since Φ is a nonlinear function, defining the linear predictive model g in $\mathcal{Z}$ is equivalent to defining a nonlinear predictive model $f : \mathcal{G} \to \mathbb{R}$, described as $f(G) = g(\Phi(G))$.

Let $g(\mathbf{z}) = \mathbf{z}^\top \boldsymbol{\varphi}^*$ be the predictive linear model $g : \mathcal{Z} \to \mathbb{R}$, where $\boldsymbol{\varphi}^* \in \mathcal{Z}$ are the optimal parameters to be estimated. Without losing generality, we consider a ridge regression in the latent space. This involves finding the best parameters by minimizing the regularized mean square error given by

$$\min_{\boldsymbol{\varphi}} \frac{1}{N} \sum_{i=1}^{N} \left(y_i - \Phi(G_i)^\top \boldsymbol{\varphi} \right)^2 + \lambda \|\boldsymbol{\varphi}\|_2^2, \tag{7}$$

where the importance of the regularization term is weighted by λ. The optimal solution vector $\boldsymbol{\varphi}^*$ for this regularized optimization problem is obtained by nullifying its gradient, resulting in

$$\boldsymbol{\varphi}^* = (\mathbf{Z}^\top \mathbf{Z} + \lambda \mathbb{I}_D)^{-1} \mathbf{Z}^\top \mathbf{y}. \tag{8}$$

where

$$\mathbf{Z} = \left(\Phi(G_1) \cdots \Phi(G_N) \right)^\top$$
$$\mathbf{y} = \left(y_1 \cdots y_N \right)^\top.$$

Then using (8) leads to the optimal predictive parameters in $\mathcal{Z}$ and we can therefore specify the nonlinear predictive model $f : \mathcal{G} \to \mathbb{R}$ by combining the transformation function Φ and the linear regression model g. Using this method, the quantitative value prediction for any graph $G \in \mathcal{G}$ is achieved by

$$f(G) = \Phi(G)^\top (\mathbf{Z}^\top \mathbf{Z} + \lambda \mathbb{I}_D)^{-1} \mathbf{Z}^\top \mathbf{y}.$$

3.3   Pre-imaging

The availability of Φ^{-1} allows the pre-image of any point of interest from the latent space to be computed, thus eliminating the pre-image problem. We propose a pre-image generation method to obtain new data given a quantitative label y. Indeed, our regression approach creates a linear relation between quantitative labels and positions in $\mathcal{Z}$. As the Gaussian distribution characterized by $(\boldsymbol{\mu}_1, \boldsymbol{\Sigma}_1)$ is intentionally associated with the minimum quantitative label in Y, and the one defined by $(\boldsymbol{\mu}_2, \boldsymbol{\Sigma}_2)$ is associated to the maximum quantitative label in Y, we can determine the position of the mean $\boldsymbol{\mu}_y$ that corresponds to a quantitative label y by employing (4). Then, from the Gaussian distribution parameterized by $(\boldsymbol{\mu}_y, \boldsymbol{\Sigma}_y)$, it is possible to sample a point $\mathbf{z} \in \mathcal{Z}$ with $\mathbf{z} \sim \mathcal{N}(\boldsymbol{\mu}_y, \sigma^2 \mathbb{I}_D)$ where σ^2 represents the variance of the Gaussian distributions chosen during the training of the model and obtain its pre-image in $\mathcal{G}$, namely

$$\hat{G} = \Phi^{-1}(\mathbf{z}).$$

4 Experiments

We evaluated our approach[1] in order to answer two separate questions:

Q1 Can the latent space produced by our regression method for graph data be considered effective, and are the representations generated suitable for regression-based objectives?

Q2 Does our model preserve its ability to efficiently generate pre-images ?

The analysis was conducted on three molecular regression datasets. In these datasets, the nodes encode the atoms of the molecule and are labeled by the chemical element of each atom. The edges encode the chemical bonds between atoms. The **QM7 dataset** is a quantum chemistry dataset composed of $7,165$ small organic molecules with up to 7 significant atoms. Its regression task consist in predicting the atomization energy of each molecule. The **ESOL** dataset is composed of $1,128$ molecular compounds, with a maximum of 55 nodes. As the used graph representation is sensitive to the size of the graphs, molecules with more than 22 atoms were filter out, thus reducing the dataset to $1,015$ different graphs. The prediction task consists of predicting the solubility measurement associated with each molecule. Finally, the **FREESOLV** dataset provides experimental and calculated information on the hydration free energies of 643 small molecules in water, with a maximum of 24 nodes. Similarly to the ESOL dataset, the dataset was reduced to 632 distinct graphs with a maximum size of 22 nodes.

We implemented our approach using **MoFlow** [20] and compared it to standard graph kernels, such as **Weisfeiler-Lehman** (WL) [18], **Shortest-Path** (SP) [4] and **Hadamard Code** (Hadcode) [12]. In addition, as our approach consists in applying ridge regression on the concatenation of the flattened representations of the graph in $\mathcal{Z}$, we also compared to simpler kernels working on the concatenation of the flattened representations of the graph in $\mathcal{G}$. We considered standard vector-based kernels: linear, RBF, Polynomial and a sigmoid [17]. Each predictive model was trained by minimizing the cost function described in (7). Finally, the capability of generating pre-images was compared with the approach outlined in [1], employed on previously mentioned kernel methods.

Each dataset was split to 90% for training and the remaining 10% for evaluation. To ensure a fair comparison, the kernels were fine-tuned and their parameters determined by cross-validation with a grid search where 10 values of λ ranging from 10^{-5} to 10^2 were selected for sampling. In addition, for the simpler kernels, a logarithmic scale was used to sample 5 weighting values applied to the similarity measure used in the kernel ranging from 10^{-5} to 10^3. The power of the Polynomial kernel was varied over a set of candidate values: $1, 2, 3, 4$. For regression evaluations, the performance is measured by the R^2 score.

As described in previous sections, we converted the graph data into a combination of a node feature matrix and an adjacency tensor, namely $G = (\mathbf{X}, \mathbf{A}) \in \mathcal{G}$ with $\mathbf{X} \in \mathbb{R}^{n \times d}$ and $\mathbf{A} \in \mathbb{R}^{n \times n \times e}$. The conducted experiments used labeled edges

---

[1] For sake of reproducibility, all experiments can be reproduced from the available GitHub repository https://github.com/clement-g28/nf-kernel.

Table 1. R^2 score ($\pm$ std) on graph regression datasets

Method		Datasets		
		QM7	ESOL	FREESOLV
Standard Kernels	Linear	0.681 ± 0.001	0.555 ± 0.032	0.254 ± 0.114
	RBF	0.680 ± 0.002	0.558 ± 0.032	0.262 ± 0.113
	Polynomial	0.681 ± 0.001	0.566 ± 0.034	0.264 ± 0.102
	Sigmoid	0.673 ± 0.002	0.563 ± 0.037	0.255 ± 0.074
Graph Kernels	WL	0.490 ± 0.0	0.602 ± 0.0	0.895 ± 0.0
	SP	0.721 ± 0.0	0.531 ± 0.0	0.543 ± 0.0
	Hadcode	0.491 ± 0.0	0.573 ± 0.0	$\mathbf{0.901} \pm 0.0$
(Ours)	MoFlow	$\mathbf{0.730} \pm \mathbf{0.008}$	$\mathbf{0.685} \pm 0.039$	0.754 ± 0.042

in each dataset, leading to the definition of e as the number of edge labels plus one label for the non-existent edge. The value of d was determined by the number of node labels along with one label for the non-existent node. In addition, since a graph composed of n nodes can be represented in $n!$ distinct ways using such a representation method, we trained our models by implementing a random permutation transformation of the input graph data. Additionally, to assess the permutation variability of our technique, all experiments were evaluated 10 times by employing random permutations. Consequently, the performance are reported as the mean performance and its standard deviation.

To answer **Q1**, Table 1 displays the average performance and standard deviation (std) on the regression datasets. These results show the good predictive performance of our method when employing linear ridge regression within the latent space $\mathcal{Z}$ for most datasets. However, best results on the FREESOLV dataset are achieved by the Weisfeiler-Lehman and Hadamard Code kernels. To understand this point, it is noteworthy that this dataset comprises a smaller number of distinct graphs, which are relatively larger in size when compared to other datasets like QM7. Moreover, the quantitative values in the FREESOLV dataset are non-uniformly distributed, ranging between -5.48 and 1.79, with a significant proportion (more than 97%) falling within the range of -2.57 to 1.79. As a result, the linear interpolation-based approach used in this study may not be the most suitable method for such datasets. However, our approach has the advantage over graph kernels that once the model has been learned, predictions can be made directly and simply while graph kernel methods can be computationally more expensive. Moreover, we can observe in the results non-zero standard deviation in our performances due to the nature of the used graph representations, as opposed to the graph kernels that are designed to be permutation invariant. Therefore, the question **Q1** can be answered positively in most cases, indicating that the nonlinear transformation Φ learned by our method is able of producing a good representation space for a regression task.

100   C. Glédel et al.

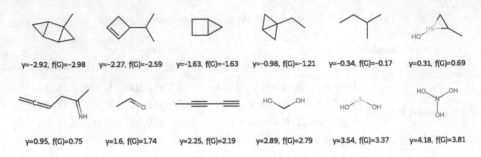

y=-2.92, f(G)=-2.98   y=-2.27, f(G)=-2.59   y=-1.63, f(G)=-1.63   y=-0.98, f(G)=-1.21   y=-0.34, f(G)=-0.17   y=0.31, f(G)=0.69

y=0.95, f(G)=0.75   y=1.6, f(G)=1.74   y=2.25, f(G)=2.19   y=2.89, f(G)=2.79   y=3.54, f(G)=3.37   y=4.18, f(G)=3.81

Fig. 2. The pre-images generated from 12 points in $\mathcal{Z}$ sampled uniformly between the $\min(Y)$ and $\max(Y)$.

To answer **Q2** and test the ability of our model to generate pre-images, we generated molecules from sampling points in $\mathcal{Z}$ using our model trained on the QM7 dataset. These points were sampled by interpolating 12 values of y between $\min(Y)$ and $\max(Y)$ as follows: For each value, we sampled a point in $\mathcal{Z}$ with $\mathbf{z} \sim \mathcal{N}(\boldsymbol{\mu}_y, \sigma^2 \mathbb{I}_D)$. Corresponding pre-images were generated in $\mathcal{X}$ by applying the inverse transformation Φ^{-1} to the generated z. Figure 2 shows the obtained pre-images, as well as the sampled quantitative values y and the predicted value $f(G)$ using our learned prediction model. This experiment demonstrates that our model can generate meaningful pre-images of points in $\mathcal{Z}$ that are not a part of the dataset, hence answering positively to **Q2**.

5   Conclusion

Our paper presented a novel approach that overcomes the curse of the pre-image using NFs for a graph regression task. Our method generates a supervised space where linear regression can be efficiently operated, as demonstrated by the conducted experiments. The results indicated that the obtained latent space is efficient in solving graph regression problems using straightforward linear operations. Moreover, the method enabled interpretability by facilitating the transformation from the latent space to the input space and generating pre-images of specific points of interest.

Our approach contributes to the application of NFs in a specific task and has the potential to be adapted for other tasks such as classification. Although our method is sensitive to the permutation of the graph due to the used representation, it may be interesting to extend it to other types of graph representations, making it possible to achieve permutation invariance.

Acknowledgements. The authors acknowledge the support of the French *Agence Nationale de la Recherche* (ANR), under grant ANR-18-CE23-0014.

References

1. Bakır, G.H., Weston, J., Schölkopf, B.: Learning to find pre-images. Adv. Neural Inf. Process. Syst. **16**, 449–456 (2004)
2. Bakır, G.H., Zien, A., Tsuda, K.: Learning to find graph pre-images. In: Rasmussen, C.E., Bülthoff, H.H., Schölkopf, B., Giese, M.A. (eds.) DAGM 2004. LNCS, vol. 3175, pp. 253–261. Springer, Heidelberg (2004). https://doi.org/10.1007/978-3-540-28649-3_31
3. Balcilar, M., Renton, G., Héroux, P., Gaüzère, B., Adam, S., Honeine, P.: Analyzing the expressive power of graph neural networks in a spectral perspective. In: International Conference on Learning Representations, Vienna, Austria (2021)
4. Borgwardt, K.M., Kriegel, H.P.: Shortest-path kernels on graphs. In: Fifth IEEE International Conference on Data Mining (ICDM 2005) (2005)
5. Bresson, X., Laurent, T.: A two-step graph convolutional decoder for molecule generation (2019). arXiv preprint arXiv:1906.03412
6. De Cao, N., Kipf, T.: Molgan: an implicit generative model for small molecular graphs (2018). arXiv preprint arXiv:1805.11973
7. Guo, X., Zhao, L.: A systematic survey on deep generative models for graph generation. IEEE Trans. Pattern Anal. Mach. Intell. **45**, 5370–5390 (2022)
8. Honda, S., Akita, H., Ishiguro, K., Nakanishi, T., Oono, K.: Graph residual flow for molecular graph generation (2019). arXiv preprint arXiv:1909.13521
9. Honeine, P., Richard, C.: Solving the pre-image problem in kernel machines: a direct method. In: 2009 IEEE International Workshop on Machine Learning for Signal Processing, pp. 1–6. IEEE (2009)
10. Jia, L., Gaüzère, B., Honeine, P.: A graph pre-image method based on graph edit distances. In: Proceedings of S+SSPR 2020 (2021)
11. Jia, L., Gaüzère, B., Honeine, P.: Graph kernels based on linear patterns: theoretical and experimental comparisons. Expert Syst. Appl. **189**, 116095 (2022)
12. Kataoka, T., Inokuchi, A.: Hadamard code graph kernels for classifying graphs. In: ICPRAM, pp. 24–32 (2016)
13. Kingma, D.P., Dhariwal, P.: Glow: Generative flow with invertible 1×1 convolutions. In: Advances in Neural Information Processing Systems, pp. 10215–10224 (2018)
14. Kobyzev, I., Prince, S.J., Brubaker, M.A.: Normalizing flows: an introduction and review of current methods. IEEE Trans. Pattern Anal. Mach. Intell. **43**(11), 3964–3979 (2020)
15. Liu, J., Kumar, A., Ba, J., Kiros, J., Swersky, K.: Graph normalizing flows. Adv. Neural Inf. Process. Syst. **32** (2019)
16. Madhawa, K., Ishiguro, K., Nakago, K., Abe, M.: Graphnvp: an invertible flow model for generating molecular graphs (2019). arXiv preprint arXiv:1905.11600
17. Schölkopf, B., Smola, A.J., Bach, F., et al.: Learning with Kernels: Support Vector Machines, Regularization, Optimization, and Beyond. MIT press, Cambridge (2002)
18. Shervashidze, N., Schweitzer, P., Van Leeuwen, E.J., Mehlhorn, K., Borgwardt, K.M.: Weisfeiler-lehman graph kernels. J. Mach. Learn. Res. **12**(9) (2011)
19. Shi, C., Xu, M., Zhu, Z., Zhang, W., Zhang, M., Tang, J.: Graphaf: a flow-based autoregressive model for molecular graph generation (2020). arXiv preprint arXiv:2001.09382
20. Zang, C., Wang, F.: Moflow: an invertible flow model for generating molecular graphs. In: Proceedings of the 26th ACM SIGKDD International Conference on Knowledge Discovery & Data Mining, pp. 617–626 (2020)

Matching-Graphs for Building Classification Ensembles

Mathias Fuchs[1]([envelope]) [iD] and Kaspar Riesen[1,2] [iD]

[1] Institute of Computer Science, University of Bern, 3012 Bern, Switzerland
{mathias.fuchs,kaspar.riesen}@unibe.ch
[2] Institute for Informations Systems, University of Applied Science and Arts
Northwestern Switzerland, 4600 Olten, Switzerland
kaspar.riesen@fhnw.ch

Abstract. Ensemble learning is a well known paradigm, which combines multiple classification models to make a final prediction. Ensemble learning often demonstrates significant benefits, in particular a better classification performance than the individual ensemble members. However, in order to work properly, ensemble methods require a certain diversity of its members. One way to increase this diversity is to randomly select a subset of the available data for each classifier during the training process (known as bagging). In the present paper we propose a novel graph-based bagging ensemble that consists of graph neural networks. The novelty of our approach is that the ensemble operates on substantially augmented graph sets. The graph augmentation technique, in turn, is based on so-called matching-graphs, which can be computed on arbitrary pairs of graphs. In an experimental evaluation on five graph data sets, we show that this novel augmentation technique paired with a bagging ensemble is able to significantly improve the classification accuracy of several reference systems.

Keywords: Graph Matching · Matching-Graphs · Graph Edit Distance · Graph Augmentation · Graph Neural Network · Ensemble Learning

1 Introduction

Graphs, which consist of nodes that might be connected by edges, are used in a wide range of applications [1]. Indeed, graphs offer a compelling alternative to vector-based approaches, especially for applications involving complex data. This is mainly because graphs are capable of encoding more information than just an ordered and fixed-size list of real numbers.

In the last four decades a large number of procedures for graph-based pattern recognition has been proposed in the literature [2]. Those procedures range from graph edit distance [3], over spectral methods [4], to graph kernels [5] (to name just three examples). Recently, with the advent of Graph Neural Networks (GNNs) [6], the power of (deep) neural networks can finally be utilized by graphs.

Supported by Swiss National Science Foundation (SNSF) Project Nr. 200021_188496.

In the present paper, we propose to use an ensemble learning method based on individual GNNs. In order to produce performant and robust GNNs, a lot of training data is typically required. Furthermore, it is accepted that ensemble learning methods perform the best when a large diversity of the individual classifiers is given [7]. The major contribution of the present paper is a novel method to generate both large and heterogeneous sets of graph data (particularly suited for ensemble learning). The novel method is based on a recently introduced data structure (known as *matching-graph* [8]).

Roughly speaking, a matching-graph encodes the matching subgraphs of two graphs under consideration. This basic concept can be used in many different ways. Matching-graphs can, for instance, improve the quality of graph dissimilarity computations by aggregating a matching-graph based distance and the original distance [8]. They can also be used to produce a subgraph based vector space embedding, by checking whether or not a set of given matching-graphs occur in the graph to embed [8]. The framework of matching-graphs is also successfully adopted for the automatic detection of relevant (i.e., frequent) substructures in very large graph sets [9]. Lastly, matching-graphs are also employed for graph augmentation in order to even out very small graph data sets [10], as well as to build more stable GNNs [11].

In the present work, we propose to further optimize the augmentation method presented in [11] to generate even more diverse matching-graphs. These novel matching-graphs provide a natural way to create realistic, diverse and relevant graphs of a specific class. It is our main hypothesis that the large amount of possible matching-graphs in conjunction with a bagging procedure ensures the diversity of the individual classifiers and finally allows to build a robust ensemble.

The remainder of this paper is organized as follows. In Sect. 2, we formally introduce the concept of matching-graphs and show how they can be used to augment a given training set of graphs and build a bagging ensemble. Eventually, in Sect. 3, we conduct an exhaustive experimental evaluation to provide empirical evidence that this novel approach is able to improve the classification accuracy of diverse reference systems. Finally, in Sect. 4, we conclude the paper and discuss potential ideas for future work.

2   Building an Ensemble with Matching-Graphs

Ensemble methods aim at combining several individual classifiers into one system. That is, an ensemble weighs the opinions of its individual members and combines their results to get the final decision [7]. Various ensemble methods have been proposed in the literature (e.g. Boosting [12] or Bagging [13]). In the present paper we employ – in principle – a bagging ensemble for graph classification. Thereby, the ensemble consists of multiple GNN classifiers that are trained on random subsets of the training data. The main contribution is to substantially increase the diversity of the bagging ensemble by means of matching-graphs. For this reason, we first introduce this basic concept (Subsect. 2.1) and then explain how these matching-graphs can be used for bagging (Subsect. 2.2).

2.1 Matching-Graphs

In the present paper, we use the following formalism to define graphs. A *graph* g is a four-tuple $g = (V, E, \mu, \nu)$, where V is the finite set of nodes, $E \subseteq V \times V$ is the set of edges, $\mu : V \rightarrow L_V$ is the node labeling function, and $\nu : E \rightarrow L_E$ is the edge labeling function.

Intuitively speaking, a matching-graph is built by extracting information about the matching of pairs of graphs and formalising this information into a new graph [8]. Matching-graphs in their original form can actually be interpreted as denoised core structures of the underlying graphs, and always refer to subgraphs of the original graphs. Therefore, to augment a given training set, the original definition of a matching-graph is not suitable. In [10], we propose an adapted version of a matching-graph that represents a mixed version of both original graphs, without being just a subgraph. However, this definition is still not optimal for the present purposes, since the resulting matching-graphs are always smaller than, or equal to, the original graphs. Hence, we now propose a further altered definition for matching-graphs more suited for the present context of increasing ensemble diversity.

The process of creating matching-graphs can be described as follows. Given a pair of graphs $g = (V, E, \mu, \nu)$ and $g' = (V', E', \mu', \nu')$, the *graph edit distance* is computed first[1]. The basic idea of graph edit distance is to transform g into g' using *edit operations* (*substitutions*, *deletions*, and *insertions* of both nodes and edges). We denote the substitution of two nodes $u \in V$ and $v \in V'$ by $(u \rightarrow v)$, the node deletion by $(u \rightarrow \varepsilon)$, and the node insertion by $(\varepsilon \rightarrow v)$, where ε refers to the empty node. By computing the graph edit distance one obtains a dissimilarity score $d(g, g')$, as well as a (sub-optimal) edit path $\lambda(g, g') = \{e_1, \ldots, e_s\}$ that consists of s edit operations that transform the source graph g in to the target graph g'.

Based on $\lambda(g, g')$ two matching-graphs $m_{g \times g'}$ and $m_{g' \times g}$ can now be built. Initially, $m_{g \times g'}$ and $m_{g' \times g}$ refer to the source graph g and the target graph g', respectively. In our procedure, we first define the partial edit path $\tau(g, g') = \{e_{(1)}, \ldots, e_{(t)}\} \subseteq \lambda(g, g')$ with $t = \lfloor p_1 \cdot s \rfloor$ edit operations, where $t < s$ is the amount of randomly selected edit operations from $\lambda(g, g')$ based on a certain probability $p_1 \in [0, 1]^2$. Next, each edit operation $e_i \in \tau(g, g')$ is applied on graphs $m_{g \times g'}$ and $m_{g' \times g}$ according to the following three rules.

Case 1. e_i is a substitution $(u \rightarrow v)$: The labels of the matching nodes $u \in V$ and $v \in V'$ are exchanged in both $m_{g \times g'}$ and $m_{g' \times g}$. Note that this operation shows no effect, if the labels of the involved nodes are identical (i.e. $\mu(u) = \mu(v)$).

Case 2. e_i is a deletion $(u \rightarrow \varepsilon)$: Node $u \in V$

– is deleted in $m_{g \times g'}$.

---

[1] For this purpose, we use algorithm BP [14] with cubic time complexity.
[2] We use here the expression p_1 (with subscript 1), because later in the paper we will introduce a second probability p_2.

– is inserted in $m_{g' \times g}$.

As we only execute parts of the edit path, it is possible that the adjacent nodes of u are not yet processed, which means that we do not know the edge structure of u in $m_{g' \times g}$. In this case, we perform a look-ahead to include edges from u to the corresponding nodes in $m_{g' \times g}$. Formally, for all node substitutions $(v \rightarrow u') \in \lambda(g, g')$, where node $v \in V$ is adjacent to node $u \in V$, we insert an edge between the inserted node u and node u' in $m_{g' \times g}$.

Case 3. e_i is an insertion $(\varepsilon \rightarrow v)$: Node $v \in V'$

– is deleted in $m_{g' \times g}$.
– is inserted in $m_{g \times g'}$ (using a similar look-ahead technique as defined for Case 2).

The basic rationale to apply these rules is that we aim at creating matching-graphs that are indeed related to the underlying graphs, but also substantially differ to them in significant ways. This is achieved by allowing both insertions of nodes and swappings of node labels.

Clearly, if p_1 is set to 1.0, $\tau(g, g')$ is equal to $\lambda(g, g')$, and thus all edit operations from the complete edit path are executed during the matching-graph creation. In this case, $m_{g' \times g}$ would be equal to the source graph g and $m_{g \times g'}$ would be equal to the target graph g'. For probabilities $p_1 < 1$, however, we obtain matching-graphs that are more diverse and particularly different from simple subgraphs (due to relabelled nodes and potential insertions). That is, when all edit operations of $\tau(g, g')$ are applied, both matching-graphs represent somehow intermediate graphs between g and g'.

Due to the flexibility of graph edit distance, the matching-graph can be built using graphs with any given labeling functions μ and ν. In other words it does not matter whether the graphs are unlabaled or contain categorical or continuous node and/or edge labels.

Figure 1 shows a visual example of an edit path $\lambda(g, g')$ between two graphs g and g' and two possible resulting matching-graphs $m_{g \times g'}$ and $m_{g' \times g}$. Both matching-graphs are created with the partial edit path that consists of $t = 3$ edit operations. In this example, it is clearly visible that neither $m_{g \times g'}$ nor $m_{g' \times g}$ is a subgraph of g or g', respectively. Furthermore, the effects of the look-ahead technique is visible. More specifically, between the inserted node $b \in V'$ and node $3 \in V$ an edge is inserted, even though the substitution $(3 \rightarrow c)$ is not yet carried out.

Note that the proposed process can lead to isolated nodes, despite look-ahead technique (for a detailed explanation of this phenomenon see [8]). As we aim to build graphs with nodes that are actually connected to at least one other node in the graph, we remove isolated nodes from the matching-graphs whenever they occur in our method.

2.2   Bagging with Matching-Graphs

Based on the process of creating matching-graphs for any pair of graphs, we can augment a given training set with virtually any number of additional graphs. In

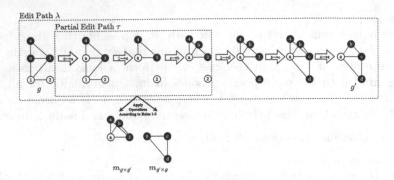

Fig. 1. An example of a complete edit path $\lambda(g, g')$, a partial edit path τ, and the resulting matching-graphs $m_{g \times g'}$ and $m_{g' \times g}$.

order to do this, we conduct the basic steps formalized in Algorithm 1 (which is similar in structure to the procedure described in [11]). The algorithm takes k sets of training graphs $G_{\omega_1}, \ldots, G_{\omega_k}$ stemming from k different classes $\omega_1, \ldots, \omega_k$, and builds two matching-graphs $m_{g \times g'}$ and $m_{g' \times g}$ for each possible graph pair $g, g' \in G_{\omega_i} \times G_{\omega_i}$. Note that the probability $p_1 \in [0.1, 0.9]$ used for the creation of the matching-graphs is randomly defined for each pair of graphs g, g' (see line 5). Assuming n training graphs per class ω_i this algorithm results in $n(n-1)$ matching-graphs, which are directly used to augment the corresponding training set G_{ω_i}. Hence, rather than n graphs, we now have access to $n(n-1) + n = n^2$ graphs per class ω_i[3].

Based on the augmented sets, a bagging ensemble $\mathcal{E} = \{c_1, \ldots, c_m\}$ with m classifiers can now be built. Each classifier $c_i \in \mathcal{E}$ is trained only on a subset of all training graphs. To this end, each classifier c_i of the ensemble $\mathcal{E}$ is trained on $\lfloor p_2 \times n^2 \rfloor$ randomly selected graphs from G_{ω_i}, where $p_2 \in [0, 1]$ is a predefined probability and n^2 is the number of graphs available in G_{ω_i} (i.e., we assume that G_{ω_i} is augmented to size $|G_{\omega_i}| = n^2$).

As base classifiers $c_i \in \mathcal{E}$, we use GNNs, which are – due to their inherent randomness – viable ensemble members. GNNs allow for the use of deep learning on graph structured data. The general goal of GNNs is to learn vector embeddings $h_v \in \mathbb{R}^n$ or $h_g \in \mathbb{R}^n$ that represent nodes $v \in V$ or complete graphs g, respectively. This vector space embedding can then be used for classification purposes. In order to learn an appropriate vector representation, GNNs implement a neighborhood aggregation strategy, called *neural message passing*, in which messages are exchanged between the nodes of a graph [15]. In the present paper, we employ a model that consists of *Graph Convolutional Layers* [16], denoted as

---

[3] By defining a further **for** loop inside the second **for** loop (in Algorithm 1 Line 5), just before the definition of p_1, even more than one matching-graph could be created for each pair of graphs, viz. we could produce more than $n(n-1)$ matching-graphs.

Algorithm 1: Graph Augmentation Algorithm

---

    input : sets of graphs from k different classes $\mathcal{G} = \{G_{\omega_1}, \ldots, G_{\omega_k}\}$
    output: same sets augmented by matching-graphs

1 **foreach** *set $G_{\omega_i} \in \mathcal{G}$* **do**
2 | $M = \{\}$
3 | **foreach** *pair of graphs $g, g' \in G_{\omega_i} \times G_{\omega_i}$* **do**
4 | | Compute $\lambda(g, g') = \{e_1, \ldots, e_s\}$
5 | | Randomly define p_1 in $[0.1, 0.9]$
6 | | Define τ by selecting $\lfloor p_1 \cdot s \rfloor$ edit operations from λ
7 | | Build both matching-graphs $m_{g \times g'}$ and $m_{g' \times g}$ according to τ
8 | | $M = M \cup \{m_{g \times g'}, m_{g' \times g}\}$
9 | **end**
10 | $G_{\omega_i} = G_{\omega_i} \cup M$
11 **end**

GCN from now on[4]. For the final graph classification, we add a dropout layer and feed the graph embedding into a fully connected layer. The outputs of the individual classifiers are then aggregated into one single decision by means of majority voting.

3 Experimental Evaluation

3.1 Data Sets and Experimental Setup

The experimental evaluation is conducted on five data sets obtained from the TUDatasets repository[5]. The first three data sets contain graphs that represent chemical compounds (*NCI1*, *PTC(MR)* and *COX-2*). The fourth data set (*Cuneiform*) contains graphs that represent Hittie cuneiform signs[6]. The last data set (*Synthie*) is an algorithmically created data set. The graphs of the first three data sets consist of nodes labeled with discrete values and unlabelled edges, whereas both *Cuneiform* and *Synthie* contain real-valued continuous node labels and unlabeled edges. Each data set is split into a training and test set according to a 4:1 split.

The novel ensemble (denoted as GCN-e_{mg}) uses the augmented training data and is built as described in Sect. 2.2. We set p_2 to 0.3 and due to computational reasons we limit the amount of selected graphs to 100'000 per class. For each ensemble we create 100 classifiers, which are trained for 200 epochs (except for the *NCI1* data set, where we build 50 classifiers, trained for 50 epochs only, due to computational problems arising from the large number of graphs in this data set).

For all base classifiers (viz. GCNs) we use the Adam optimizer with an initial learning rate of 0.01, together with a CosineAnnealingLR scheduler. Furthermore, we use the Cross Entropy loss function. The batch size is set to 64. For

---

[4] Any other classifier could be used for the experiments, as long as both the reference ensemble and our novel ensemble are based on the same classifier.

[5] https://graphlearning.io.

[6] One of the oldest handwriting systems in the world.

the implementation of the ensemble we use the ensemble-pytorch library[7] which we adapted to seamlessly work with PyTorch Geometric [17][8].

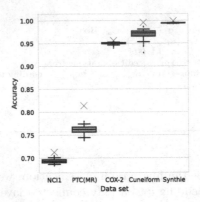

Fig. 2. Training accuracies of all individual classifiers of the ensemble for all data sets shown with a box-plot. The training accuracy of the ensemble is marked with a red cross. (Color figure online)

Figure 2 shows by means of box-plots the training accuracies of all individual classifiers available in the ensembles for all data sets. The diamonds above and below the boxes mark the 10% best and worst classifiers w.r.t. the accuracy. The training accuracy of the final ensemble is marked with a red cross. We observe that the diversity of the classifiers is the largest for NCI1, PTC(MR), and Cuneiform. It is also clearly visible that for all data sets the training accuracy of the complete ensemble is better than the accuracy of the best individual member. This is already a clear indication for the usefulness of the defined ensemble.

3.2   Reference Systems

The overall aim of the present experimental evaluation is to answer the question, whether or not matching-graphs can be beneficially employed to build robust classifier ensembles. In order to answer this research question, we use three reference systems for comparisons with our novel approach GCN-e$_{mg}$.

- *Reference system 1* (denoted by GCN): This reference system is trained on the full training set to obtain a baseline for the classification accuracy. In order to counteract uncontrolled randomness during initialization, each experiment that uses this reference system is repeated five times and the average accuracy is finally reported.

---

[7] https://ensemble-pytorch.readthedocs.io.
[8] https://pytorch-geometric.readthedocs.io.

We also perform an ablation study in order to empirically verify that the superiority of the proposed method is indeed based on the matching-graphs. To this end, we compare our novel ensemble GCN-e_{mg} with two additional reference systems.

- *Reference system 2* (denoted by GCN-e): This reference system is virtually the same as our novel ensemble but has only access to the original training data without matching-graphs.
- *Reference system 3* (denoted by SINGL-e_{mg}): This reference system refers to the best individual classifier of the novel augmented ensemble.

A comparison with reference system 2 allows us to better assess whether the matching-graphs, or the ensemble by itself, is the important element of the whole process. A comparison with reference system 3 allows us to assess whether the ensemble outperforms the randomly generated members of the system – in other words, whether the ensemble actually also makes a difference.

3.3   Test Results and Discussion

In Table 1 we compare the novel ensemble GCN-e_{mg} with the first reference system (a single GCN trained on the full training set). Remember that we run the GCN reference system five times to counteract randomness during initialization. This is why we report here the mean accuracies ($\pm$ standard deviation).

We observe that GCN-e_{mg} outperforms the reference system GCN in 18 out of 25 cases with statistical significance[9]. On the NCI1 data set, even though we observe an improvement in all five iterations, only four of them are statistically significant. On the PTC(MR) data set, we outperform the reference system in each iteration, however only two of the improvements are statistically significant. On the COX-2 data set we get an improvement in four out of five iterations (two of them are actually statistically significant). On Cuneiform and Synthie all of the improvements are statistically significant (10 out of 10 cases).

Table 1. Average classification accuracy of reference system 1 (GCN) compared to our novel ensemble (GCN-e_{mg}). Symbol $\widehat{x}$ indicates a statistically significant improvement in x out of the five comparisons (using a Z-test at significance level $\alpha = 0.05$). Marked in bold is the best accuracy per data set.

	Ref. System 1	Ours
Dataset	GCN	GCN-e_{mg}
NCI1	70.5 ± 1.0	**74.0** ④
PTC(MR)	62.3 ± 4.5	**68.6** ②
COX-2	70.4 ± 11.1	**78.7** ②
Cuneiform	40.3 ± 20.5	**96.7** ⑤
Synthie	76.8 ± 7.5	**97.5** ⑤

---

[9] The statistical significance is computed via Z-test at significance level $\alpha = 0.05$.

Table 2. Classification accuracy of the ensemble without matching-graphs GCN-e, the individually best classifier SINGL-e$_{mg}$ and our ensemble with matching-graphs GCN-e$_{mg}$. Symbols ∘/∘ indicate a statistically significant improvement over the second/third reference system using a Z-test at significance level $\alpha = 0.05$). Marked in bold is the best accuracy per data set.

Dataset	Ref. System 2 GCN-e	Ref. System 3 SINGL-e$_{mg}$	Ours GCN-e$_{mg}$
NCI1	70.7	**74.8**	74.0 ∘/–
PTC(MR)	61.4	62.9	**68.6** ∘/∘
COX-2	73.4	77.6	**78.5** –/–
Cuneiform	68.3	93.3	**96.7** ∘/–
Synthie	64.2	93.8	**97.5** ∘/–

Next, in Table 2 we compare the novel ensemble with the other two reference systems (for the sake of an ablation study). First, we observe that the best single classifier of each ensemble (reference system 3) outperforms the baseline ensemble (reference system 2) on all data sets. This is a clear indication of the usefulness of the matching-graphs. Even more important, the proposed ensemble GCN-e$_{mg}$ outperforms the second reference ensemble GCN-e on all data sets. These improvements are statistically significant on four out of five data sets. This is a strong indication that the matching-graphs are the important factor in improving the classification accuracy, rather than primarily the ensemble itself. However, when comparing our ensemble with the third reference system, it is also obvious that the ensemble still makes an important contribution – only on NCI1 is the best individual classifier better than the ensemble.

4   Conclusion and Future Work

Ensemble learning is often able to improve the accuracy of single classification systems. One popular way to build an ensemble is bagging, which combines the output of many classifiers into one strong prediction. One of the main problems in building a robust ensemble is that large and diverse data sets are needed. In the present work we propose to use so-called matching-graphs to substantially increase the amount of training data available. On the basis of these augmented training sets of graphs, a classifier ensemble is built via bagging. As base classifiers for the ensemble we use GNNs (note, however, that any other classifier could be used as well).

By means of an experimental evaluation, we empirically confirm that our novel approach significantly outperforms three related reference systems, viz. a single GNN classifier, a bagging ensemble trained on the original training set, as well as the best individual classifier stemming from the novel ensemble. Hence, we conclude that matching-graphs provide a versatile way to generate large sets of additional graphs in order to build a diverse and robust ensemble.

For future work we see at least two rewarding avenues that can be pursued. First, we could explore other ensemble modalities (rather than bagging), and second, other aggregation techniques to combine the results could be explored (rather than majority voting).

References

1. Foggia, P., Percannella, G., Vento, M.: Graph matching and learning in pattern recognition in the last 10 years. Int. J. Pattern Recogn. Artif. Intell. **28**(1), 1450001 (2014). https://doi.org/10.1142/S0218001414500013
2. Conte, D., Foggia, P., Sansone, C., Vento, M.: Thirty years of graph matching in pattern recognition. Int. J. Pattern Recogn. Artif. Intell. **18**(3), 265–298 (2004). https://doi.org/10.1142/S0218001404003228
3. Gao, X., Xiao, B., Tao, D., Li, X.: A survey of graph edit distance. Pattern Anal. Appl. **13**(1), 113–129 (2010). https://doi.org/10.1007/s10044-008-0141-y
4. Kang, U., Hebert, M., Park, S.: Fast and scalable approximate spectral graph matching for correspondence problems. Inf. Sci. **220**, 306–318 (2013). https://doi.org/10.1016/j.ins.2012.07.008
5. Kriege, N.M., Johansson, F.D., Morris, C.: A survey on graph kernels. Appl. Netw. Sci. **5**(1), 6 (2020). https://doi.org/10.1007/s41109-019-0195-3
6. Scarselli, F., Gori, M., Tsoi, A.C., Hagenbuchner, M., Monfardini, G.: The graph neural network model. IEEE Trans. Neural Netw. **20**(1), 61–80 (2009). https://doi.org/10.1109/TNN.2008.2005605
7. Dong, X., Yu, Z., Cao, W., Shi, Y., Ma, Q.: A survey on ensemble learning. Front. Comput. Sci. **14**(2), 241–258 (2020). https://doi.org/10.1007/s11704-019-8208-z
8. Fuchs, M., Riesen, K.: A novel way to formalize stable graph cores by using matching-graphs. Pattern Recogn. **131**, 108846 (2022). https://doi.org/10.1016/j.patcog.2022.108846
9. Fuchs, M., Riesen, K.: Iterative creation of matching-graphs - finding relevant substructures in graph sets. In: Proceedings of the 25th Iberoamerican Congress on Pattern Recognition, CIARP25 2021 (2021)
10. Fuchs, M., Riesen, K.: Graph augmentation for small training sets using matching-graphs. In: ICPRAI - 3rd International Conference on pattern Recognition and Artificial Intelligence (2022)
11. Fuchs, M., Riesen, K.: Graph augmentation for neural networks using matching-graphs. In: Gayar, N.E., Trentin, E., Ravanelli, M., Abbas, H. (eds.) Artificial Neural Networks in Pattern Recognition - 10th IAPR TC3 Workshop, ANNPR 2022, Proceedings. Lecture Notes in Computer Science, Dubai, United Arab Emirates, 24–26 November 2022, vol. 13739, pp. 3–15. Springer, Heidelberg (2022). https://doi.org/10.1007/978-3-031-20650-4_1
12. Hastie, T., Rosset, S., Zhu, J., Zou, H.: Multi-class adaboost. Stat. Interface **2**(3), 349–360 (2009)
13. Breiman, L.: Bagging predictors. Mach. Learn. **24**(2), 123–140 (1996). https://doi.org/10.1007/BF00058655
14. Riesen, K., Bunke, H.: Approximate graph edit distance computation by means of bipartite graph matching. Image Vis. Comput. **27**(7), 950–959 (2009). https://doi.org/10.1016/j.imavis.2008.04.004
15. Gilmer, J., Schoenholz, S.S., Riley, P.F., Vinyals, O., Dahl, G.E.: Neural message passing for quantum chemistry. In: International Conference on Machine Learning, pp. 1263–1272. PMLR (2017)

16. Kipf, T.N., Welling, M.: Semi-supervised classification with graph convolutional networks. In: 5th International Conference on Learning Representations, ICLR 2017, Conference Track Proceedings, Toulon, France, 24–26 April 2017. OpenReview.net (2017). https://openreview.net/forum?id=SJU4ayYgl
17. Fey, M., Lenssen, J.E.: Fast graph representation learning with PyTorch Geometric. In: ICLR Workshop on Representation Learning on Graphs and Manifolds (2019)

Maximal Independent Sets for Pooling in Graph Neural Networks

Stevan Stanovic[1]([✉]) [iD], Benoit Gaüzère[2] [iD], and Luc Brun[1] [iD]

[1] Normandie Univ, ENSICAEN, CNRS, UNICAEN, GREYC UMR 6072,
14000 Caen, France
{stevan.stanovic,luc.brun}@ensicaen.fr
[2] INSA Rouen Normandie, Univ Rouen Normandie, Université Le Havre Normandie,
Normandie Univ, LITIS UR 4108, 76000 Rouen, France
benoit.gauzere@insa-rouen.fr

Abstract. Convolutional Neural Networks (CNNs) have enabled major advances in image classification through convolution and pooling. In particular, image pooling transforms a connected discrete lattice into a reduced lattice with the same connectivity and allows reduction functions to consider all pixels in an image. However, there is no pooling that satisfies these properties for graphs. In fact, traditional graph pooling methods suffer from at least one of the following drawbacks: Graph disconnection or overconnection, low decimation ratio, and deletion of large parts of graphs. In this paper, we present three pooling methods based on the notion of maximal independent sets that avoid these pitfalls. Our experimental results confirm the relevance of maximal independent set constraints for graph pooling.

Keywords: Graph Neural Networks · Graph Pooling · Graph Classification · Maximal Independent Set · Edge Selection

1 Introduction

Convolutional Neural Networks (CNNs) achieved major advances in computer vision by learning abstract representations of images thought convolution and pooling. A convolution is a linear filter applied to each pixel of an image which combines its value with the one of its surrounding. The resulting value is usually transformed via a non linear function. The pooling step reduces the size of an image by grouping a connected set of pixels, usually a small squared window, in a single pixel whose value is computed from the ones of window's pixel. Graph Neural Networks (GNNs) take their inspiration from CNNs and aim at transferring advances performed on images to graphs. However, most of CNNs use images with a fixed structure (shape). While using GNN both the structure of a graph and its content varies from one graph to another. Convolution and pooling operations must thus be adapted for graphs.

A GNN may be defined as a sequence of simple graphs $(\mathcal{G}^{(0)}, \ldots, \mathcal{G}^{(m)})$ where each $\mathcal{G}^{(l)} = (\mathcal{V}^{(l)}, \mathcal{E}^{(l)})$ is produced by layer l from $\mathcal{G}^{(l-1)}$. Sets $\mathcal{V}^{(l)}$ and $\mathcal{E}^{(l)}$ denote respectively the set of vertices and the set of edges of the graph. Given $n_l = |\mathcal{V}^{(l)}|$, the

M. Vento et al. (Eds.): GbRPR 2023, LNCS 14121, pp. 113–124, 2023.
https://doi.org/10.1007/978-3-031-42795-4_11

graph $\mathcal{G}^{(l)}$ may be alternatively defined as $\mathcal{G}^{(l)} = (\mathbf{A}^{(l)}, \mathbf{X}^{(l)})$ where $\mathbf{A}^{(l)} \in \mathbb{R}^{n_l \times n_l}$ is the weighted adjacency matrix of $\mathcal{G}^{(l)}$ while $\mathbf{X}^{(l)} \in \mathbb{R}^{n_l \times f_l}$ encodes the nodes' attributes of $\mathcal{G}^{(l)}$ whose dimension is denoted by f_l. Each line u of $\mathbf{X}^{(l)}$ encodes the feature of the vertex u and is denoted by $x_u^{(l)}$.

The final graph $G^{(m)}$ of a GNN is usually followed by a Multi-Layer Perceptron (MLP) applied on each vertex for a node prediction task or by a global pooling followed by a MLP for a global graph classification task.

Graph convolution. This operation is mainly realized by a message passing mechanism and allows to learn a new representation for each node by combining the information of the mentioned node and its neighborhood. The neighborhood information is obtained by aggregating all the adjacent nodes information. Therefore, the message passing mechanism can be expressed as follows [8]:

$$\mathbf{x}_u^{(l+1)} = UPDATE^{(l)}(\mathbf{x}_u^{(l)}, AGGREGATE^{(l)}(\{\mathbf{x}_v^{(l)}, \forall v \in \mathcal{N}(u)\})) \qquad (1)$$

where $\mathcal{N}(u)$ is the neighborhood of u and $UPDATE, AGGREGATE$ correspond to differentiable functions.

Let us note that convolution operations should be permutation equivariant, i.e. for any permutation matrix $P \in \{0, 1\}^{n_l \times n_l}$ defined at level l, if f denotes the convolution defined at this layer we must have: $f(PX^{(l)}) = Pf(X^{(l)})$. Note that this last equation, together with Eq. 1, hides the matrix $\mathbf{A}^{(l)}$ which nevertheless plays a key role in the definition of the $AGGREGATE$ function by defining the neighborhood of each node.

Global pooling. For graph level tasks, a fixed size vector needs to be sent to the MLP. However, due to the variable sizes of graphs within a dataset, global pooling must aggregate the whole graph information into a fixed size vector. This operation can be performed by basic operators like sum, mean or maximum. Let note us that more complex aggregation strategies [19] also exist. To insure that two isomorphic graphs have the same representation, global pooling must be invariant to permutations, i.e. for any permutation matrix P, defined at layer l we must have $g(PX^{(l)}) = g(X^{(l)})$ where g denotes the global pooling operation.

Hierarchical pooling. Summing up a complex graph into a fixed size vector leads necessarily to an important loss of information. The basic idea to attenuate this loss consists in gradually decreasing the size of the input graph thanks to pooling steps inserted between convolution layers. The resulting smaller final graph induces a reduced loss of information in the final global pooling step. This type of method is called a hierarchical pooling [12, 18]. The hierarchical pooling step, as the convolution operation should be permutation equivariant in order to keep information localised on desired nodes. Conversely, global pooling must be permutation invariant since it computes a graph level representation. Let note that, similar to CNNs, the reduced graph leads to a reduction of parameters in the next convolution. However, this reduction is mitigated by the learned part of hierarchical pooling. Moreover, let us consider a line graph with a signal optimally sampled on its vertices. As shown by [2], most of GNN correspond to a low pass filter. Applying a GNN on this line graph, hence decreases the maximal frequency of our signal on vertices producing an over sampling according to the Nyquist theorem. More details on optimal sampling on graphs may be found in [1, 15].

Given a graph $\mathcal{G}^{(l)} = (\mathbf{A}^{(l)}, \mathbf{X}^{(l)})$ defined at layer l and its reduced version $\mathcal{G}^{(l+1)} = (\mathbf{A}^{(l+1)}, \mathbf{X}^{(l+1)})$ defined at level $l + 1$, the connection between $\mathcal{G}^{(l)}$ and $\mathcal{G}^{(l+1)}$ is usually insured by the reduction matrix $\mathbf{S}^{(l)} \in \mathbb{R}^{n_l \times n_{l+1}}$ where n_l and n_{l+1} denote respectively the sizes of $\mathcal{G}^{(l)}$ and $\mathcal{G}^{(l+1)}$. If $\mathbf{S}^{(l)}$ is a binary matrix, each column of $\mathbf{S}^{(l)}$ encodes the vertices of $\mathcal{G}^{(l)}$ which are merged into a single vertex at layer $l + 1$. If $\mathbf{S}^{(l)}$ is real, each line of $\mathbf{S}^{(l)}$ encodes the distribution of each vertex of $\mathcal{G}^{(l)}$ over the vertices of $\mathcal{G}^{(l+1)}$. In both cases, we require $\mathbf{S}^{(l)}$ to be line-stochastic.

Given $\mathcal{G}^{(l)} = (\mathbf{A}^{(l)}, \mathbf{X}^{(l)})$ and $\mathbf{S}^{(l)}$, the feature matrix $\mathbf{X}^{(l+1)}$ of $\mathcal{G}^{(l+1)}$ is defined as follows:

$$X^{(l+1)} = S^{(l)\top} X^{(l)} \tag{2}$$

This last equation defines the attribute of each surviving vertex v_i as a weighted sum of the attributes of the vertices v_j of $\mathcal{G}^{(l)}$ such that $\mathbf{S}_{ji}^{(l)} \neq 0$.

The adjacency matrix of $\mathcal{G}^{(l+1)}$ is defines by:

$$A^{(l+1)} = S^{(l)\top} A^{(l)} S^{(l)} \tag{3}$$

Let us suppose that $\mathbf{S}^{(l)}$ is a binary matrix. Each entry (i, j) of $\mathbf{A}^{(l+1)}$ defined by Eq. 3 is equal to $\sum_{r,s}^{n_l} \mathbf{A}_{r,s}^{(l)} \mathbf{S}_{r,i}^{(l)} \mathbf{S}_{s,j}^{(l)}$. Hence two surviving vertices i and j are adjacent in $\mathcal{G}^{(l+1)}$ if it exists at least two adjacent non surviving vertices r and s such that r is merged onto i ($\mathbf{S}_{r,i}^{(l)} = 1$) and s onto j($\mathbf{S}_{s,j}^{(l)} = 1$).

Pooling Methods. There are two main families of pooling methods. The first family, called Top-k methods [7, 12], is based on a selection of relevant vertices based on a learned criteria. The second family is based on node's clustering methods as in Diff-Pool [18].

Top-k methods such as gPool [7] learn a score attached to each vertex by computing the scalar product between the vertex's attributes and one or several learned vectors. Alternatively, a GNN can be used to compute a relevance vector for each vertex as in SagPool [12]. Next, a fixed ratio pooling is used to select the k vertices with a highest score. Unselected vertices are dropped. In this case, two surviving vertices in the reduced graph will be adjacent only if they were adjacent before the reduction. This last point may result in the creation of disconnected reduced graphs. This disconnection may be avoided by increasing the density of the graph, using power 2 or 3 of its adjacency matrix or by using the Kron's reduction [3] instead of Eq. 3. Nevertheless, let us note that simply discarding all non surviving vertices leads to an important loss of information. We proposed in a previous contribution [14], a top-k pooling method called MIVSPool which avoids such drawbacks by using a maximal independent vertex set and graph contraction operations.

Clustering based methods learn explicitly or implicitly the matrix $\mathbf{S}^{(l)}$ which encodes the reduction of a set of vertices at level l into a single vertex at level $l+1$. Methods (eg. [18]) learning $\mathbf{S}^{(l)}$ explicitly have to use a predetermined number of clusters. This last point forbids the use of graphs of different sizes. Additionally, these methods generally result in dense matrices $\mathbf{S}^{(l)}$ which then induce dense adjacency matrices at

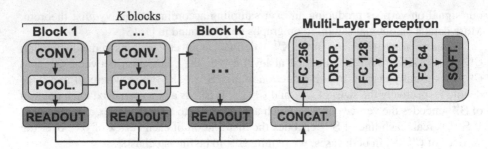

Fig. 1. General architecture of our GNN. Each block is composed of a convolution layer followed by a pooling layer. Features learned after each block are sent to the next block and a Readout layer. The K vectors resulting from each Readout are concatenated to have several levels of description of the graph and, finally, the concatenation is sent to a Multi-Layer Perceptron.

level $l + 1$ (Eq. 3). As a consequence, graphs produced by these pooling methods have a density close to 1 (i.e. a complete graph or an almost complete graph).

An alternative strategy consists in learning $\mathbf{S}^{(l)}$ only implicitly. Graph pooling such as the maximal matching method used in EdgePool [4] may be associated to this strategy. A maximal matching of a graph $\mathcal{G}^{(l)} = (\mathcal{V}^{(l)}, \mathcal{E}^{(l)})$ is a subset M of $\mathcal{E}^{(l)}$, where no two edges are incident to a same vertex, and every edge in $\mathcal{E}^{(l)} \setminus M$ is incident to one of the two endpoints of an edge in M. EdgePool is based on a maximal weighted matching technique, i.e. a maximal matching of maximal weight. The weight of each edge, called its score, is learned using the attributes of its two end points. The selected edges are then contracted to form a specific cluster. Note that the use of a maximal weighted matching may result in some vertices not incident to any selected edges. These vertices are left unchanged. The sequential algorithm [4] has been parallelized by Landolfi [11]. Unlike EdgePool, Landolfi [11] learns a score attached to each vertex and sort all the vertices of the graph according to their score. The weight of each edge is then defined from a combination of the rank of its incident nodes. The similarity between two adjacent vertices is in this case not taken into account. Moreover, both EdgePool and Landolfi [11] have a decimation ratio lower than 50%, which suggests the need for more pooling steps or a poor abstraction in the final graph of the GNN.

In this paper, we propose an unified family of graph pooling methods which maintains a decimation ratio of approximately 50%, while simultaneously preserving both the structure of the original graph and its attribute information. We achieve this by using a Maximal Independent Set (MIS) [9] to select surviving edges that are evenly distributed throughout the graph, and by assigning non-surviving elements to those that do survive. As a result, we avoid any subsampling or oversampling issues that may arise (see Fig. 2). The source code of the paper is available on the CodeGNN ANR Project Git repository: https://scm.univ-tours.fr/projetspublics/lifat/codegnn.

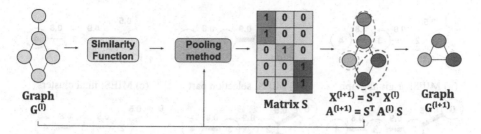

Fig. 2. General proposition of our three graph poolings. Each edge is associated to a similarity score (Sect. 2). Based on this similarity, a MIS on edge is computed from which a reduction matrix S is derived. Applying S to both feature and structure leads to a reduced graph $G^{(l+1)}$.

2 Maximal Independent Sets and Graph Poolings

2.1 Maximal Independent Set (MIS) and Meer's Algorithm

Definition. Let $\mathcal{X}$ be a finite set and $\mathcal{N}$ a neighborhood function defined on $\mathcal{X}$ such that the neighborhood of each element includes the element itself. A subset $\mathcal{J}$ of $\mathcal{X}$ is a Maximal Independent Set (MIS) if the two following equations are fulfilled:

$$\forall (x, y) \in \mathcal{J}^2 : x \notin \mathcal{N}(y) \tag{4}$$

$$\forall x \in \mathcal{X} - \mathcal{J}, \exists y \in \mathcal{J} : x \in \mathcal{N}(y) \tag{5}$$

The elements of $\mathcal{J}$ are called the surviving elements or survivors. Equations (4) and (5) respectively states that two surviving elements can't be neighbors and each non-surviving element has to be in the neighborhood of at least one element of $\mathcal{J}$. These two equations can be interpreted as a subsampling operation where Eq. (4) is a condition preventing the oversampling (two adjacent vertices cannot be selected) while Eq. (5) prevents subsampling: Any non-surviving element is at a distance 1 from a surviving one.

A way to compute a MIS is the Meer's algorithm [13] which only involves local computations and is therefore parallelizable. This algorithm attaches a variable to each element. Let us denote by $\mathcal{J}$ the current maximal independent set at an iteration of the algorithm, and let us additionally consider the value v_x attached to an element x. Then x is added to $\mathcal{J}$ at current iteration if v_x is maximal among the values of $\mathcal{N}(x) - \mathcal{N}(\mathcal{J})$, where $\mathcal{N}(\mathcal{J})$ denotes $\mathcal{J}$ and its neighbors. Meer's algorithm provides thus a maximal matching such that each of its element is a local maxima at a given step of the algorithm. We can thus interpret the resulting set as a maximal weight independent set.

Assignment of Non-surviving Elements. After a MIS, $\mathcal{X}$ is split in two subsets: the surviving elements contained in the set $\mathcal{J}$ and the non-surviving elements contained in $\mathcal{X} - \mathcal{J}$. Simply considering $\mathcal{J}$ as a digest of X may correspond to an important loss of information which simply discards $\mathcal{X} - \mathcal{J}$. In order to avoid such a loss we allow each non surviving element contained in $\mathcal{X} - \mathcal{J}$ to transfer its information to a survivor. The possibility of such a transfer is insured thanks to Eq. 5 which states that each non

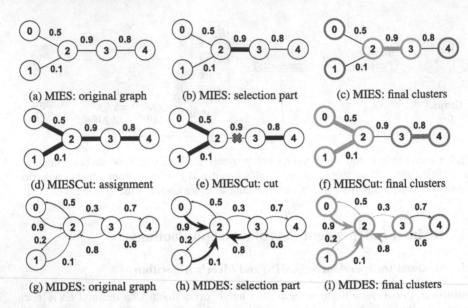

(a) MIES: original graph   (b) MIES: selection part   (c) MIES: final clusters

(d) MIESCut: assignment   (e) MIESCut: cut   (f) MIESCut: final clusters

(g) MIDES: original graph   (h) MIDES: selection part   (i) MIDES: final clusters

Fig. 3. Schema of our proposed methods on a toy graph. Number on each edge corresponds to its score s and the bold edges indicates the surviving ones. Each group of vertices with the same color represent a cluster. Figures 3a and 3b are common steps for MIES and MIESCut.

surviving element is adjacent to at least one survivor. We can thus associate to any non surviving element x_j a surviving neighbor denoted by $\sigma(x_j)$. At layer l, the global assignment of non-surviving elements onto surviving ones is encoded by the reduction matrix $\mathbf{S}^{(l)} \in \mathbb{R}^{n_l \times n_{l+1}}$ such that :

$$\mathbf{S}_{ii}^{(l)} = 1 \quad \forall x_i \in \mathcal{J} \text{ and } \mathbf{S}_{j\sigma(j)}^{(l)} = 1 \quad \forall x_j \in \mathcal{X} - \mathcal{J} \tag{6}$$

with $\mathbf{S}_{ij}^{(l)} = 0$ otherwise.

2.2 Maximal Independent Sets for Graph Pooling

Based on the work [9] defined within the image partitioning framework we introduce in the following, three adaptations of these methods in order to define learnable pooling steps. In the following sections, the adjacency matrix $\mathbf{A}^{(l+1)}$ is obtained from $\mathbf{A}^{(l)}$ and a binary version of $\mathbf{S}^{(l)}$ using Eq. 3.

Maximal Independent Edge Set. Most of pooling methods are based on a projection score for each vertex. This strategy is based on the assumption that we can learn features characterizing relevant vertices for a given classification task. However, even if this hypothesis holds, two adjacent vertices may have similar scores and the choice of the survivor is in this case arbitrary. An alternative strategy consists in merging similar nodes. Given a GNN with hierarchical pooling, the graph sequence corresponds to an increasing abstraction from the initial graphs. Consequently, vertices encoded at each layer of the GNN encode different types of information. Based on this observation, we decided to learn a similarity measure between adjacent vertices at each layer. Inspired by [16], we define the similarity at layer l between two adjacent vertices u and v as $s_{uv}^{(l)} = exp(-\|\mathbf{W}^{(l)}.(x_u - x_v)\|)$ where x_u and x_v are the features of vertices u and v, $\mathbf{W}^{(l)}$ is a learnable matrix and $\|.\|$ is the L_2 norm.

Given the maximal weighted matching $\mathcal{J}^{(l)}$ defined at level l, each vertex of $\mathcal{G}^{(l)}$ is incident to at most one edge of $\mathcal{J}^{(l)}$. If $u \in \mathcal{V}^{(l)}$ is not incident to $\mathcal{J}^{(l)}$ its features are just duplicated at the next layer. Otherwise, u is incident to one edge $e_{uv} \in \mathcal{J}^{(l)}$ and both u and v are contracted at the next layer. Since u and v are supposed to be similar the attributes of the vertex encoding the contraction of u and v at the next layer must be symmetrical according to u and v. To do so, we first define the attribute of e_{uv} as

$$x_{uv} = \frac{1}{2}(x_u^{(l)} + x_v^{(l)}) \tag{7}$$

where x_u and x_v are the features of vertices u and v. The attribute of the merged vertex is then defined as $s_{uv}x_{uv}$.

An equivalent update of the attributes of the reduced graph may be obtained by computing the matrix $\mathbf{S}^{(l)}$ encoding the transformation from $\mathcal{G}^{(l)}$ to $\mathcal{G}^{(l+1)}$. This matrix can be defined as $\mathbf{S}_{ii}^{(l)} = 1$ if i is not incident to $\mathcal{J}^{(l)}$, and by selecting arbitrary one survivor among $\{u, v\}$ if $e_{uv} \in \mathcal{J}^{(l)}$. If u is selected we set $\mathbf{S}_{uu}^{(l)} = \mathbf{S}_{vu}^{(l)} = \frac{1}{2}s_{uv}$. All remaining entries of $\mathbf{S}^{(l)}$ are set to 0. Matrix $\mathbf{X}^{(l+1)}$ can then be obtained using Eq. 2. We call this method MIESPool and the main steps are presented in Figs. 3a to 3c.

Maximal Independent Edge Set with Cut (MIESCut). Graph reduction through maximal weighted matching has two main drawbacks within the GNN framework. First, a maximal matching may produce many vertices not adjacent to the set of selected edges. Such vertices are just copied to the next level which induce a low decimation ratio (lower than 50%). Given that, the number of layers of a GNN is usually fixed, this last drawback may produce a graph with an insufficient level of abstraction at the final layer of the GNN. Secondly, only the score of the selected edges are used to compute the reduced attributes. This last point reduces the number of scores used for the back-propagation and hence the quality of the learned similarity measures. As in the previous section, let us denote by $\mathcal{J}^{(l)}$ the maximal weighted matching defined at layer l. By definition of a maximal weighted matching, each vertex not incident to $\mathcal{J}^{(l)}$ is adjacent to at least one vertex which is incident to $\mathcal{J}^{(l)}$. Following [9], we increase the decimation ratio, by attaching isolated vertices to contracted ones. This operation is performed by selecting for each isolated vertex u the edge e_{uv} such that s_{uv} is maximal and v is incident to $\mathcal{J}^{(l)}$.

This operation provides a spanning forest of $\mathcal{G}^{(l)}$ composed of isolated edges, trees of depth one (called stars) with one central vertex and paths of length 3. This last type of tree corresponds to a sequence of 4 vertices with strong similarities between any pair of adjacent vertices along the paths. However, merging all 4 vertices into a single one, suppose implicitly to apply twice an hypothesis on the transitivity of our similarity measure: more precisely the fact that the two extremities of the paths are similar is not explicitly encoded by our selection of edges. In order to avoid such assumption we remove the central edge of such paths from the selection in order to obtain two isolated edges (see Figs. 3d to 3f).

Let us denote by $\mathcal{J}'^{(l)}$ the resulting set of selected edges which forms a spanning forest of $\mathcal{G}^{(l)}$ composed of isolated edges and stars. Concerning the definition of $\mathbf{S}^{(l)}$, we apply the same procedure than in the previous section for isolated edges. For stars, we select the central vertex as the surviving vertex. Let us denote by u such a star's center. We then set $\mathbf{S}_{uu}^{(l)} = \frac{1}{2}$ and $\mathbf{S}_{vu}^{(l)} = \frac{1}{2M}s_{uv}$ for any v such that $e_{uv} \in \mathcal{J}'^{(l)}$ where M is a normalizing factor defined as: $M = \sum_{v|e_{uv} \in \mathcal{J}'^{(l)}} s_{uv}$. The computation of the attributes of the reduced graph using Eq. 2 and matrix $\mathbf{S}^{(l)}$ is equivalent to compute for each star's center u, the sum, weighted by the score, of the edges' attributes (Eq. 7) incident to u and belonging to $\mathcal{J}'^{(l)}$:

$$x_u^{(l+1)} = \frac{1}{\sum_{v|e_{uv} \in \mathcal{J}'^l} s_{uv}} \sum_{v|e_{uv} \in \mathcal{J}'^l} s_{uv} x_{uv}^{(l)} \tag{8}$$

Maximal Independent Directed Edge Set. The definition of a spanning forest composed of isolated edges and stars is obtained in three steps by MIESCut: The definition of a maximal weight matching, the attachment of isolated vertices and the cut of all paths of length 3. Following [9], we propose to use the Maximal Independent Directed Edge set (MIDES) reduction scheme which obtains the same type of spanning forest in a single step. This reduction scheme is based on a decomposition of the edges e_{uv} of the undirected graphs in two oriented edges $e_{u \to v}$ and $e_{v \to u}$. The neighborhood of an oriented edge $\mathcal{N}(e_{u \to v})$ is defined as the union of the sets of edges leaving u, arriving on u and leaving v. Given $\mathcal{G}^{(l)}$ defined at layer l we formally have:

$$\mathcal{N}^{(l)}(e_{u \to v}) = \{e_{u \to v'} \in \mathcal{E}^{(l)}\} \cup \{e_{v' \to u} \in \mathcal{E}^{(l)}\} \cup \{e_{v \to v'} \in \mathcal{E}^{(l)}\} \tag{9}$$

The main difference between the neighborhoods defined by Eq. 9 and the one of MIES is that we do not include in the neighborhood edges arriving on v. This asymmetry allows the creation of stars centered on v. The MIDES algorithm computes a MIS on the edge set using the neighborhood defined by (9) (see Fig. 3g to 3i).

At layer l, applying a MIDES on $\mathcal{G}^{(l)}$ requires to define a score function on directed edges. We propose to use $s_{uv} = exp(-\|W.(x_u - x_v) + b\|)$ where the bias term b allows to introduce an asymmetry so that $s_{uv} \neq s_{vu}$ if $x_u \neq x_v$.

Let us denote by $\mathcal{D}^{(l)}$ the set of directed edges produced by a MIDES on $\mathcal{G}^{(l)}$ using our scoring function. The set $\mathcal{D}^{(l)}$ defines on $\mathcal{G}^{(l)}$ a spanning forest composed of isolated vertices, isolated edges and stars [9].

For an isolated vertex u we duplicate this vertex at the next layer and copy its attributes. We thus set $\mathbf{S}_{uu}^{(l)} = 1$.

For an isolated directed edge $e_{u \to v} \in \mathcal{D}^{(l)}$ we select v as a surviving vertex and set $\mathbf{S}_{vv}^{(l)} = \frac{s_{uv}}{M}$ and $\mathbf{S}_{uv}^{(l)} = \frac{s_{vu}}{M}$ where $M = s_{uv} + s_{vu}$. This setting corresponds to the following update of the attributes: $x_v^{(l+1)} = (s_{uv}.x_v^{(l)} + s_{vu}.x_u^{(l)})/(s_{uv} + s_{vu})$. Let us note that since $e_{u \to v} \in \mathcal{D}^{(l)}$ we have $s_{uv} > s_{vu}$. The previous formula put thus more weight on the surviving vertex v. This update may be considered as a generalization of Eq. 7 using the asymmetric scores s_{uv} and s_{vu}.

A star within the MIDES framework is defined by a set of edges $e_{w \to v}$ of $\mathcal{D}^{(l)}$ arriving on the same vertex v. We then set v as survivor and generalize the update of the attributes defined for isolated edges by setting $\mathbf{S}_{vv}^{(l)} = \frac{1}{N} \sum_{u | e_{u \to v} \in \mathcal{D}^{(l)}} \frac{s_{uv}}{M_u}$ and $\mathbf{S}_{uv}^{(l)} = \frac{1}{N} \frac{s_{vu}}{M_u}$ for all u such that $e_{u \to v} \in \mathcal{D}^{(l)}$ where $M_u = s_{uv} + s_{vu}$ and N is the number of such u. Such a definition of $\mathbf{S}^{(l)}$ is equivalent to set the updated attribute of v as the mean value of its incident selected edges:

$$
x_v^{(l+1)} = \frac{1}{N} \sum_{u | e_{u \to v} \in \mathcal{D}^{(l)}} \frac{s_{uv} x_v^{(l)} + s_{vu} x_u^{(l)}}{s_{uv} + s_{vu}} \text{ with } N = |\{ u \in \mathcal{V}^{(l)} | e_{u \to v} \in \mathcal{D}^{(l)} |.
$$

3 Experiments

Datasets. We evaluate our contribution to a bio-informatics and a social dataset called respectively D&D [5] and REDDIT-BINARY [17] whose statistics are reported on Table 1. The aim of D&D is to classify proteins as either enzyme or non-enzyme. Nodes represent the amino acids and two nodes are connected by an edge if they are less than 6 Ångström apart. REDDIT-BINARY is composed of graphs corresponding to online discussions on Reddit. In each graph, nodes represent users, and there is an edge between them if at least one of them respond to the other's comment. A graph is labeled according to whether it belongs to a question/answer-based community or a discussion-based community.

Model Architecture and Training Procedure. Our model architecture is composed of K blocks where each block consists of a GCN [10] convolution layer followed by a pooling layer. The vector resulting of each pooling operation is then sent to the next block (if it exists) and a Readout layer. A Readout layer concatenates the average and the maximum of vertices' features matrix $\mathbf{X}^{(l)}$ and these K concatenations are themselves concatenated and sent to a Multi-Layer Perceptron (MLP). The MLP is composed of three fully connected layers and a dropout is applied between each of them. Finally, a Softmax layer is used to determine the binary class of graphs. Note that no batch normalization is applied (Fig. 1).

To evaluate our model, we use the training procedure proposed by [6]. This procedure performs an outer 10-fold cross-validation (CV) to split the dataset into ten training

Table 1. Statistics of datasets

Dataset	#Graphs	#Classes	Avg. $|\mathcal{V}|$	Avg. $|\mathcal{E}|$
D&D [5]	1178	2	284 ± 272	715 ± 694
REDDIT-BINARY [17]	2000	2	430 ± 554	498 ± 623

Table 2. Average classification accuracies obtained by different pooling methods. Highest and second highest accuracies are respectively in **bold** and blue. $\pm$ indicates the 95% confidence interval of classification accuracy.

Methods	D&D [5]	REDDIT-BINARY [17]
Baseline	76.29 ± 2.33	87.07 ± 4.72
gPool [7]	75.61 ± 2.74	84.37 ± 7.82
SagPool [12]	76.15 ± 2.88	85.63 ± 6.26
EdgePool [4]	72.59 ± 3.59	87.25 ± 4.78
MIVSPool [14]	76.35 ± 2.09	$\mathbf{88.73 \pm 4.43}$
MIESPool	77.17 ± 2.33	88.08 ± 4.55
MIESCutPool	$\mathbf{77.74 \pm 2.85}$	86.47 ± 4.57
MIDESPool	76.52 ± 2.21	88.40 ± 4.74

and test sets. For each outer fold, another 10-fold CV (inner) is applied to the training set to select the best hyperparameter configuration. Concerning hyperparameters, learning rate is set to 10^{-3}, weight decay to 10^{-4} and batch size to 512. Other hyperparameters are tuned using a grid search to find the best configuration. Possible values for the hidden layers sizes are $\{64, 128\}$, dropout ratio is chosen within $\{0.2, 0.5\}$ and the number of blocks K between 1 and 5. We use the Adam optimizer and maximal number of epochs is set to 1000 with an early stopping strategy if the validation loss has not been improved 100 epochs. For EdgePool, due to time constraints, we fixed the hidden layers sizes at 128 and the dropout ratio at 0.5.

We compare, in Table 2, our methods to five state-of-art methods: Baseline (K blocks of GCN [10]), gPool [7], SagPool [12], EdgePool [4] and MIVSPool [14], our previous MIS method. First, we note that the baseline obtains quite good results while not implementing any pooling strategy. It highlights the fact that defining a good pooling operation is not trivial. State-of-the-art methods mostly fail at this task, certainly due to the significant loss of information resulting from the hard selection of surviving vertices using a top$-k$ strategy. This hypothesis is confirmed by the better results obtained by MIVSPool. Let us note also that for D& D, based on T-tests with a significance level of 5%, the average accuracy of EdgePool is statistically lower than the ones of MIS methods. Second, we can observe that the strategies combining edge selection methods and MIS (MIESPool, MIESCutPool, MIDESPool) achieve either the highest or the second highest performances. This empirical results tend to demonstrate that the selection on edges may be most relevant, and that a MIS strategy improves the effectiveness of the pooling over EdgePool. Finally, best results are obtained by different MIS strategies, hence indicating that the right MIS strategy may be dataset dependant. This hypothesis has to be tested using more extensive hyperparameters selection.

4   Conclusion

Graph poolings based on Maximal Independent Sets (MIS) allow, unlike state-of-art methods, to maintain a fixed decimation ratio close to 50%, to preserve vertex information and to avoid subsampling and oversampling. Results obtained by our three methods based on MIS confirm the interest of this approach but further investigations on other datasets are needed to conclude on the effectiveness of our methods. The design of alternative similarity scores also corresponds to a promising line of research.

Acknowledgments. The work reported in this paper was supported by French ANR grant #ANR-21-CE23-0025 CoDeGNN and was performed using HPC resources from GENCI-IDRIS (Grant 2022-AD011013595) and computing resources of CRIANN (Grant 2022001, Normandy, France).

References

1. Anis, A., Gadde, A., Ortega, A.: Towards a sampling theorem for signals on arbitrary graphs. In: 2014 IEEE International Conference on Acoustics, Speech and Signal Processing (ICASSP), pp. 3864–3868. IEEE (2014)
2. Balcilar, M., Guillaume, R., Héroux, P., Gaüzère, B., Adam, S., Honeine, P.: Analyzing the expressive power of graph neural networks in a spectral perspective. In: Proceedings of the International Conference on Learning Representations (ICLR) (2021)
3. Bianchi, F.M., Grattarola, D., Livi, L., Alippi, C.: Hierarchical representation learning in graph neural networks with node decimation pooling. IEEE Trans. Neural Netw. Learn. Syst. **33**(5), 2195–2207 (2022)
4. Diehl, F., Brunner, T., Le, M.T., Knoll, A.: Towards graph pooling by edge contraction. In: ICML 2019 Workshop on Learning and Reasoning with Graph-Structured Data (2019)
5. Dobson, P.D., Doig, A.J.: Distinguishing enzyme structures from non-enzymes without alignments. J. Molecul. Biol. **330**(4), 771–783 (2003)
6. Errica, F., Podda, M., Bacciu, D., Micheli, A.: A fair comparison of graph neural networks for graph classification. arXiv preprint arXiv:1912.09893 (2019)
7. Gao, H., Ji, S.: Graph u-nets. In: International Conference on Machine Learning, pp. 2083–2092. PMLR (2019)
8. Hamilton, W.L.: Graph representation learning. Synth. Lect. Artif. Intell. Mach. Learn. **14**(3), 1–159 (2020)
9. Haxhimusa, Y.: The Structurally Optimal Dual Graph Pyramid and Its Application in Image Partitioning, vol. 308. IOS Press (2007)
10. Kipf, T.N., Welling, M.: Semi-supervised classification with graph convolutional networks. In: International Conference on Learning Representations (ICLR) (2017)
11. Landolfi, F.: Revisiting edge pooling in graph neural networks. In: ESANN (2022)
12. Lee, J., Lee, I., Kang, J.: Self-attention graph pooling. In: International Conference on Machine Learning, pp. 3734–3743. PMLR (2019)
13. Meer, P.: Stochastic image pyramids. Comput. Vis. Graph. Image Process. **45**(3), 269–294 (1989)
14. Stanovic, S., Gaüzère, B., Brun, L.: Maximal independent vertex set applied to graph pooling. In: Structural, Syntactic, and Statistical Pattern Recognition: Joint IAPR International Workshops, S+ SSPR 2022, Montreal, 26–27 August 2022, Proceedings, pp. 11–21. Springer, Cham (2023). https://doi.org/10.1007/978-3-031-23028-8_2
15. Tanaka, Y., Eldar, Y.C., Ortega, A., Cheung, G.: Sampling signals on graphs: from theory to applications. IEEE Signal Process. Magaz. **37**(6), 14–30 (2020)
16. Verma, N., Boyer, E., Verbeek, J.: Feastnet: feature-steered graph convolutions for 3d shape analysis. In: Proceedings of the IEEE Conference on Computer Vision and Pattern Recognition, pp. 2598–2606 (2018)

17. Yanardag, P., Vishwanathan, S.: Deep graph kernels. In: Proceedings of the 21th ACM SIGKDD International Conference on Knowledge Discovery and Data Mining, pp. 1365–1374 (2015)
18. Ying, Z., You, J., Morris, C., Ren, X., Hamilton, W., Leskovec, J.: Hierarchical graph representation learning with differentiable pooling. Adv. Neural Inf. Process. Syst. **31**, 4805–4815 (2018)
19. Zhang, M., Cui, Z., Neumann, M., Chen, Y.: An end-to-end deep learning architecture for graph classification. Proc. AAAI Conf. Artif. Intell. **32**(1), 4438–4445 (2018)

Graph-Based Representations
and Applications

Detecting Abnormal Communication Patterns in IoT Networks Using Graph Neural Networks

Vincenzo Carletti$^{(\boxtimes)}$, Pasquale Foggia, and Mario Vento

Department of Information Engineering, Electrical Engineering and Applied
Mathematics, University of Salerno, Fisciano, Italy
{vcarletti,pfoggia,mvento}@unisa.it
https://mivia.unisa.it

Abstract. Nowadays, millions of Internet of Things (IoT) devices communicate over the Internet, thus becoming potential targets for cyberattacks. Due to the limited hardware capabilities of these devices, host-based countermeasures are unlikely to be deployed on them, making network traffic analysis the only reasonable way to detect malicious activities. In this paper, we face the problem of identifying abnormal communications in IoT networks using graph-based anomaly detection methods. Although anomaly detection has already been applied to graph-based data, most existing methods have been used for static graphs, with the aim of detecting anomalous nodes. In our case, the graphs represent snapshots of the network traffic, and change with time. In this paper we compare different graph-based methods, and different graph representations of the network traffic, using two large datasets of real IoT data.

Keywords: Network Anomaly Detection · IoT Networks · Graph Neural Networks

1 Introduction

The Internet of Things (IoT) has revolutionized different fields leading the development of smart homes, smart medical devices, smart cities and smart industries. IoT devices are embedded with sensors and communication technologies that enable them to collect and transmit data over the internet. However, the wide diffusion of such devices, together with their limited security capabilities, has made them the preferred target of malicious users for performing Distributed Denial of Service attacks (DDoS) [15]. Common host-based countermeasures, like intrusion detection systems (HIDS) or antivirus software are unlikely to be installed on IoT devices, due to their limited hardware capabilities. Therefore, the only reasonable way to detect infected devices or malicious activities is to analyze the network exchanges observed by a dedicated monitoring device. Unfortunately, IoT network communications are highly dynamic and heterogeneous, if compared with traditional personal computer networks, making more challenging for traditional network-based intrusion detection systems (NIDS) to detect malicious traffic.

M. Vento et al. (Eds.): GbRPR 2023, LNCS 14121, pp. 127–138, 2023.
https://doi.org/10.1007/978-3-031-42795-4_12

Several methods for network analysis, based on machine learning and deep learning, have been proposed [1,10,12,13], but most of them assume that the characteristics of malicious traffic are known in advance (or, at least, can be inferred from a set of training examples). Of course, this assumption does not hold for many real-world scenarios, where we are not aware in advance of all the possible threats, and it is inpractical to collect a sufficient amount of samples of malicious traffic.

An alternative is to deal with this problem as an anomaly detection task, where the system learns a model of normal traffic, so as to classify anything that does not fit this model as a threat [6]. Anomaly detection methods for IoT networks have been reviewed and compared in some recent papers [3,6,8]. Most of them work with vectors of measurements extracted from *network flows*, i.e. sequences of network packets sharing the same transport protocol and endpoints. Although some network flows statistics have been proved to be very effective to detect anomalies in the communication between two network nodes, they seem to be less discriminant for abnormal communication patterns involving larger groups of devices [9], as in the case of botnets.

Graph-based approaches can help to overcome this limitation: a graph is a natural representation for data where the *structure*, i.e. the way the different pieces are interconnected, plays an important role.

In the scientific literature, different Graph Neural Networks (GNNs) have been proposed to detect anomalous nodes in attributed graphs [11,18]. Among these, we have selected three different approaches recently proposed, for which the authors publicly provide an implementation. DOMINANT (Deep Anomaly Detection on Attributed Networks) [5], a deep graph autoencoder where the encoder function aims to map node features to a lower-dimensional latent space, and the decoder function aims to reconstruct both the original feature matrix and the graph topology using these compressed node representations. CONAD (Contrastive Anomaly Detection) proposed in [18] to identify anomalous nodes in attributed graphs by utilizing prior knowledge. Together with the GNN the authors have also proposed a new approach to graph data augmentation that explicitly incorporates human knowledge of various anomaly types as contrastive samples. The latter correspond to nodes whose structural and semantic information deviate considerably from those of existing nodes in the input attributed network. The contrastive samples are used to generate an adversarial version of the input graph. OCGNN (One Class Graph Neural Network), a graph version of the One Class Support Vector Machine OCSVM, proposed by Wang et al. in [17]. The method aims at learning the minimum volume hyper-sphere that contains all the embeddings of normal nodes.

These graph-based methods face the detection of abnormal communication patterns as a node classification problem and are trained in a transductive context, that involves training and testing the network on a single large attributed graph. Although this is a very common approach, it does not allow to take into account the dynamic nature of the IoT networks over time. To face this limitation, in this paper, we have partitioned the network communication data into temporal snapshots as proposed in [21]. Each snapshot is represented as a distinct graph.

For the graph representation of a snapshot, we have considered three approaches: *similarity graphs* [4], that provide a general method to transform time series into graphs; *traffic trajectory graphs* [19,20], a graph representation specifically devised for network flows; and *extended TDG* [21], an extension of Traffic Dispersion Graphs [7] (TDGs) originally presented to model the behavior of network hosts. The overall network traffic is thus represented through a sequence of graphs, having a topology that can vary according to the communication flows in the corresponding temporal snapshot.

For each of the three graph-based representations, we have adapted the three mentioned GNN anomaly detection methods, modifying their learning approach from transductive (i.e. the GNN learns from different subsets of the same graph) to inductive (i.e. each training sample is a different, complete graph). The 9 resulting combinations of graph representation and GNN method have been experimentally evaluated using two large, recent datasets of real IoT traffic.

The paper is organized as follows. In Sect. 2, we describe the graph-based representations for network traffic snapshots. In Sect. 3, we provide some details on the three GNN considered. In Sect. 4, we describe the experimental setup, the used datasets and how we have trained the GNN; then we present, compare and discuss the results obtained by each combination of GNN and graph representation. Finally, in Sect. 5 we draw our conclusions about the effectiveness of using GNN for the task at hand.

2   Representing Network Traffic as a Graph

Although representing the static topology of a network as a graph is a relatively straightforward process, in which hosts are assigned to nodes and physical links to edges, there is not a unique way for representing network communications as graphs. As introduced in Sect. 1, we have selected three recent graph-based representations that are suitable for the task at hand.

Starting from the captured raw packets, we have extracted the communication flows and the related feature vectors using NFStream [2], a publicly available tool for traffic analysis. Each feature vector represents a flow, i.e. a sequence of packets having the same transport-level endpoints; the vectors contain 77 features commonly used to represent flows in network analysis, including categorical features like IP and MAC address, statistical features such as average packet size or number of packets as well as temporal features like duration and starting time.

During the extraction, in order to represent the evolution of the communication over time, as proposed in [21], we have marked each network flow with the timestamp obtained from its first captured packet and have partitioned the overall network communication in time windows of duration T; so that all the flows having the timestamp that falls in the time window $(t, t + T)$ have been grouped in the temporal snapshots $\mathtt{snap}_t$.

The feature vectors of each snapshot are the input data for building the graphs, according to the methods described in the following subsections.

2.1 Similarity Graphs

Given the snapshot snap_t, a *similarity graph* [4] is made by assigning each feature vector x_i to a different node; thus nodes do not represent network devices, but communication flows.

In the next step, a node similarity matrix is obtained by computing for each pair of nodes the cosine distance between the vectors associated to the nodes. Finally, for each node, K directed edges are added connecting it to its K closest neighbors with respect to the similarity matrix.

Although similarity graphs are a general way to represent time series as graphs, and they have not been developed to model network traffic, the rationale behind their use in our application is that the flows corresponding to normal traffic should form large, densely connected clusters of nodes, while the anomalous flows will likely become nodes that are more isolated from the rest of the graph.

2.2 Traffic Trajectory Graphs

As in similarity graphs, also in *traffic trajectory graphs* [19,20] each node represents a communication flow. However, in this case, the (undirected) edges represent the fact that two flows share one of their endpoints network-level address. In this way, if there is a device that is very active in the network, the nodes corresponding to its flows will have a lot of connections. Thus, the graph structure implicitly encodes the activity level of different parts of the network in each snapshot.

2.3 Extended Traffic Dispersion Graphs

In *Traffic Dispersion Graphs* (TDGs), the nodes are associated to the transport-level endpoints that communicate over the network, i.e. the different IP address/transport port number pairs between which communication flows are exchanged, and the (undirected) edges represet the flows; notice that the flow feature vectors thus become attributes of the graph edges, instead of the graph nodes.

Extended TDGs [21] enrich the description of a node by adding, as node attributes, both some graph-related properties such as degree, centrality, betweeness, closeness, eccentricity, and the arithmetic mean of the flow feature vectors of edges adjacent to the node.

3   Graph Neural Networks for Anomaly Detection

In this Section we provide some details about the structure and loss functions of the GNNs mentioned in Sect. 1; we will also describe how they can be used to distinguish between normal and abnormal nodes.

3.1 DOMINANT

DOMINANT is a graph autoencoder for attributed graphs composed of a *attributed graph encoder*, a *feature reconstruction decoder*, and a *topology reconstruction decoder*. The encoder part uses a GNN with three layers, each of them followed by a ReLU activation function, that compute the embedding matrix Z, associating to each node a latent vector. More formally, given a weight matrix W, the adjacency matrix A and a feature matrix H (associating a feature vector to each node of the graph), each layer computes the following function:

$$f(H, A|W) = ReLU(\hat{D}^{-\frac{1}{2}} \hat{A} \hat{D}^{-\frac{1}{2}} HW) \tag{1}$$

where $\hat{A} = A + I$ and $\hat{D}$ is the diagonal matrix with $\hat{D}_{ii} = \sum_j \hat{A}_{ij}$. The overall computation of the encoder, given the feature matrix X corresponding to the input features of the nodes, is:

$$\begin{aligned} H^{(1)} &= f(X, A|W^{(0)}) \\ H^{(2)} &= f(H^{(1)}, A|W^{(1)}) \\ Z &= f(H^{(2)}, A|W^{(2)}) \end{aligned} \tag{2}$$

At the end of this process, the latent vectors associated to the nodes (the rows of matrix Z) do not depend only on the corresponding feature vectors, but also contain information from the k-hops neighborhood of each node.

The topology decoder aims at reconstructing the matrix A from Z, using the following computation:

$$A^* = sigmoid(ZZ^T) \tag{3}$$

while the feature decoder tries to reconstruct the feature matrix X from Z, using a structure that is similar to a single layer of the encoder:

$$X^* = f(Z, A|W^{(d)}) \tag{4}$$

The loss function is a linear combination of the reconstruction errors for the topology and the features, which are measured using the Frobenius norm of the differences between the input matrices and the reconstructed ones:

$$L = (1 - \alpha)||A - A^*||_F + \alpha||X - X^*||_F \tag{5}$$

Once the network has been trained, it is used as an anomaly detector by encoding and then reconstructing the input graph. The reconstruction error is used to compute an anomaly score and rank the nodes that are most anomalous in the network; thus the method does not return whether a node is anomalous or not but only the score assigned to each node. To our purpose, we added a final thresholding layer to classify normal and abnormal nodes. The best threshold has been selected by using the (Receiver Operating Characteristic) computed on the validation set.

3.2 OCGNN

OCGNN aims at learning together an embedding for the nodes of the graph, and a hyper-sphere (in the node embedding space) that contains all the normal nodes. The method does not depend on the choice of a particular embedding function; however, the authors suggest to use one or more layers organized as we have previously seen in Eq. 1.

The learnable weights of the embedding function $g(X, A|W)$ are represented by the matrix W (X and A are, as in the previous subsection, the node features and the adjacency matrix). The hyper-sphere is represented by its center c and its radius r. However, the center is computed simply as the average of the embedding vectors of the training nodes, so only r is actually learned.

The algorithm uses as its loss function:

$$L(r, W) = \frac{1}{\beta N} \sum_{v \in V_{tr}} \left[||g(X, A|W)_v - c||^2 - r^2 \right]^+ + r^2 + \frac{\lambda}{2} \sum_{i=1}^{K} ||W^k||^2 \qquad (6)$$

where V_{tr} is the set of training nodes having cardinality N, the notation $g(\cdot)_v$ represents the row corresponding to node v of the embedding matrix computed by $g(\cdot)$, and $[\cdot]^+$ denotes a maximum operation between zero and its argument. Finally, λ is a regularization hyper-parameter while $\beta \in]0, 1[$ is a hyper-parameter used to balance the trade-off between enclosing all the embeddings in the hyper-sphere and getting the smallest radius: with the given loss function, the algorithm will try to enlarge the hyper-sphere if it contains less than a fraction β of the training samples, and to reduce it otherwise.

OCGNN uses the distance between a node embedding and the center of the hyper-sphere as a metric to provide an anomaly score. A node v is considered as anomalous if and only if $||g(X, A|W)_v - c||^2 \geq r^2$.

3.3 CONAD

CONAD is a graph autoencoder that uses data augmentation and a Siamese architecture to learn an optimal latent encoding. More specifically, given a training graph G (containing only normal nodes), CONAD generates an augmented graph G_{ano} containing artificially generated anomalous nodes. While the method does not depend on how these nodes are generated, the authors consider the following strategies: *(a)* adding a large number of random edges to a node, yielding a degree significantly higher than the average; *(b)* removing most of the connections of a node, making it mostly isolated from other nodes; *(c)* randomly modifying node features so as to be very dissimilar from their immediate neighbors; *(d)* randomly modifying node features so as to be significantly larger or significantly smaller that the values in most other nodes.

The encoder of the network is a Graph Attention Network (GAT) trained using a Siamese configuration, in which two instances of the encoder are applied on both G and G_{ano}; a contrastive loss function is used to make the corresponding embeddings obtained on the two graphs more similar to each other for the normal nodes, and more dissimilar if the node in G_{ano} was an artificially generated anomalous node:

$$L_{sc} = \frac{1}{N} \sum_{v \in V_{tr}} I_{y_v=0} \cdot ||z_v - \hat{z}_v|| + I_{y_v=1} \cdot [m - ||z_v, \hat{z}_v||]^+ \qquad (7)$$

where z_v and $\hat{z}_v$ are the encodings computed for node v in G and in G_{ano} respectively, and $I_{y_v=1}$ and $I_{y_v=0}$ are the indicator functions representing the fact that node v in G_{ano} is normal or it has been modified to become anomalous, respectively. The value m is a hyper-parameter denoting the desired *margin* between the distances for normal nodes and the distances between a normal and an anomalous node.

Similarly to DOMINANT, the CONAD network has two decoders to reconstruct both the adjacency matrix A and the feature matrix X from the encoding matrix Z. Thus, the complete loss function used for training the network is a linear combination of the contrastive loss function presented above and the two reconstruction errors. Finally, once the network has been trained, the latter is used to decide if a node is anomalous, in a way similar to DOMINANT.

4 Experiments

In this section we describe in details the experiments conducted and the obtained results.

4.1 Datasets

The experiments have been conducted using two recent publicly available datasets that provide raw network captures.

IoT23 [14] is a dataset containing benign and malicious IoT network traffic divided into 23 scenarios. Three of them consist of benign traffic captured from real IoT devices in a smart home environment. The authors then created 20 malicious scenarios by uploading different attack instances to a RaspberryPI present in the environment. The captures last 24 h, except in cases where the number of generated packets increased too rapidly.

The second dataset is IoTID20 [16], that also considers devices of a typical smart home scenario, together with smartphones and laptops. In addition, also smartphones and laptops were connected to the network during This dataset includes 9 kinds of attacks (some of them are not in IoT23), such as various categories of DoS and DDoS attacks, ARP spoofing and operating system scanning.

4.2 Graph Neural Network Training

As previously mentioned, the GNN methods we have considered for anomaly detection, were originally employed by their authors in a *transductive* setting: there is a single graph, whose structure is known a priori, and the method is used to predict the class of some nodes (i.e. whether they are normal or anomalous) given a disjoint set of nodes which are known to be normal.

In our case, that setting is not appropriate: each graph correspond to a snapshot of observed network traffic, and snapshots captured at different times will very likely have a different structure. Thus, we had to modify the training procedure to a more conventional inductive setting, where the network learns a function that is independent of the graph structure, and then this function is used to classify the nodes of new graphs, that are different from the ones seen at training time. Notice that the conventional technique used for inductive learning, i.e. applying the optimization algorithm to fixed-size batches randomly sampled from the training set, was not directly applicable here: the GNN layers and the loss functions (as described in Sect. 3) assume that the structure of the graph (the adjacency matrix A) is provided, and all the nodes of the graph are given in input. So, we have used a different, entire graph from the training set at each learning step; the training set was made of graphs consisting only of normal nodes, while the validation set also contains abnormal nodes. The training and validation procedure is repeted different times by randomly changing the order of the input graphs sequence in order to prevent the GNNs from specialyzing on a particular order of graphs.

In more details, the dataset IoT23 has been used to train and validate the GNNs, for each representation 10,041 graphs have been used for the training set and 9,227 for the validation, while 9,221 graphs have composed the test set. The graphs have been extracted from approximately 1,400,000 flows of which 1,000,000 are benign and 200,000 malicious. The dataset IoTID20 has be used only for testing the capability of the GNNs to generalize on data collect from a similar IoT context, but with different devices and network attacks. In particular, from IoTID20 we have extracted 209 graphs for each representation, starting from 108,983 malicious and 14,753 benign flows. For both the datasets the duration of a temporal snapshot used to generate the graphs has been set to 2.5 min. This choice has been mostly driven by memory constraints during the training since we needed to load the whole graphs on the GPU.

For the sake of clarity, we provide some details about the hyperparameters and the training parameters for each considered method. Regarding DOMINANT, we used an encoder with three layers and a feature reconstruction decoder with one layer. The hidden layers of the encoder had 28 nodes, and the dropout probability was 0.3. During loss and anomaly score computation, we set the value of α to 0.8 to balance the features reconstruction error and the structural reconstruction error. We optimized the GNN using ADAM with a learning rate of 0.0005. The autoencoder was trained for 100 epochs, and we set the patience level to 20. For OCGNN, we opted for a 2-layer GraphSAGE encoder, as it is known to work effectively in an inductive context. The hidden layer consists of 28 neurons. We set the contamination factor $1-\beta$, which denotes the proportion of nodes allowed to stay outside of the hyper-sphere, to 0.2. For optimization of the encoder, we used the AdamW optimizer with a learning rate of 0.0003 and decay coefficient. The model was trained for 50 epochs with a patience of 10. Finally, in the case of CONAD, we set the number of encoder layers to 3, while the number of decoder layers responsible for reconstructing features is set to 1. The hidden encoder layer is had 28 neurons, while the output layer had 14. We set the loss margin m to 0.5, and we introduced a percentage of artificial anomalies in the augmented graphs equal to 0.2% of the total number of nodes in the input graph. We used ADAM as optimizer, with a learning rate of 0.0005. The autoencoder was trained for 50 epochs with a patience of 10.

4.3 Results

In Table 1 we have reported precision, recall and F-score for all the combinations of GNN and representations previously mentioned. Since our system is an anomaly detector, we considered as positive the anomalous nodes (e.g. a true positive was a node representing an attack that was labeled as anomalous by the system).

From Table 1 it is possible to note that among the three representations, the extended TDG is performing significantly worse than the others on IoT23, the dataset used to train the GNNs. On the other hand, the results on IoTD20 may lead to a different conclusion, indeed, this representation seems to provide a good generalization capability to GNNs. But, considering the compostion of the two dataset, where the percentage of abnormal traffic in IoTD20 (about 87%) is considerably higher than in IoT23 (about 29%), we can reasonably conclude that the GNNs are specilized on IoT23 resulting in a large number of false positives on IoTID20. An evidence of that is the very high recall achieved by all the GNNs on IoTD20, while the precision is close to percentage of malicious flows. Unfortunately, this is a common problem when an anomaly detection system trained on a given scenario is moved on a different one. This is the only considered representation where graph nodes represent communication endpoints instead of communication flows. Also, note that extended TDGs include a lot of network-related features in the nodes, but they fail to bring any benefit to the detection task.

The two remaining representations, similarity graphs and traffic trajectory graphs, have similar performance, but the first one is slightly better in general. Remember that in similarity graphs, node adjacency is determined by the similarity of the communication flow characteristics, while in traffic trajectory graphs it depends only on the sharing of a common endpoint. Only while using OCGNN, we get a significant drop of recall, that can be blamed to difficulty in adapting this GNN to face the problem at hand. We suspect that similarity graphs are better suited at capturing the occurrence of similar anomalous behavior by several devices that are not immediately connected to each other, and this may help for many kinds of attacks.

Looking at the GNNs, the one that achieved the best performance in the average is DOMINANT. In particular, while working with similarity graphs, DOMINANT shows the best F-score on both the datasets, therefore it has the best trade-off between precision and recall. The results suggest that a classical graph autoencoder architecture, according to which DOMINANT has been

136   V. Carletti et al.

Table 1. Performance of the considered GNN and representations, in terms of precision (Prec), recall (Rec) and F-Score, evaluated over the test sets of both IoT23 and IoTID20 datasets.

GNN	Representation	IoT23			IoTID20		
		Prec	Rec	F-Score	Prec	Rec	F-Score
CONAD	Extended	0.4815	0.5542	0.5153	0.8502	0.9953	0.9171
	Similarity	**0.6807**	0.6725	0.6766	0.8699	0.8433	0.8564
	Trajectory	0.6756	0.7010	0.6881	0.8652	0.8441	0.8545
DOMINANT	Extended	0.4937	0.5399	0.5158	0.8466	**0.9955**	0.9168
	Similarity	0.6748	0.7324	**0.7024**	**0.8896**	0.9951	**0.9394**
	Trajectory	0.6137	**0.7814**	0.6875	0.8735	0.8855	0.8794
OCGNN	Extended	0.4857	1.0000	0.6539	0.8491	1.000	0.9184
	Similarity	0.5963	0.7268	0.6551	0.7154	0.2026	0.3158
	Trajectory	0.5000	1.0000	0.6667	0.8823	1.000	0.9375

designed, is more suitable for the task at hand. CONAD also is based on a graph autoencoder, but its use of manually designed models of anomalies introduces a bias in the system, that may lead to a decrease in performance.

Finally, OCGNN seem to be slightly behind the other two methods. OCGNN does not try to optimize reconstruction error as the other two methods do, but attempts to project the nodes into a space where the normal nodes reside in a possibly small hyper-sphere. The fact that on the test set this network shows a large difference between precision and recall appears to be a consequence of the fact that the optimal position and radius of this hyper-sphere may change when the graph structure changes, and thus this algorithm is less performant in an inductive setting than it would be in its original transductive usage.

5   Conclusions

In conclusions, in this paper we have faced the problem of detecting abnormal communication patterns in IoT networks using Graph Neural Networks. To this purpose, we have selected from the state of the art three different ways of representing network traffic in terms of graphs, and three graph-based anomaly detection algorithms. The resulting 9 combinations have been experimentally evaluated using two recent datasets composed of real IoT network traffic. proposed in the context of network analysis. The results that we have obtained in the experiments are encouraging. Further analysis is required to assess the reliability and robustness as well as the generalization capability of GNN in this context.

References

1. Abbasi, M., Shahraki, A., Taherkordi, A.: Deep learning for network traffic monitoring and analysis (NTMA): a survey. Comput. Commun. **170**, 19–41 (2021). https://doi.org/10.1016/j.comcom.2021.01.021
2. Aouini, Z., Pekar, A.: Nfstream: a flexible network data analysis framework. Comput. Netw. **204**, 108719 (2022)
3. Churcher, A., et al.: An experimental analysis of attack classification using machine learning in IOT networks. Sensors **21**(2), 446 (2021)
4. Deng, A., Hooi, B.: Graph neural network-based anomaly detection in multivariate time series. In: Proceedings of the AAAI Conference on Artificial Intelligence, vol. 35, pp. 4027–4035 (2021)
5. Ding, K., Li, J., Bhanushali, R., Liu, H.: Deep anomaly detection on attributed networks. In: Proceedings of the 2019 SIAM International Conference on Data Mining, pp. 594–602. SIAM (2019)
6. Fahim, M., Sillitti, A.: Anomaly detection, analysis and prediction techniques in IOT environment: a systematic literature review. IEEE Access **7**, 81664–81681 (2019). https://doi.org/10.1109/ACCESS.2019.2921912
7. Iliofotou, M., Pappu, P., Faloutsos, M., Mitzenmacher, M., Singh, S., Varghese, G.: Network monitoring using traffic dispersion graphs (TDGs). In: Proceedings of the 7th ACM SIGCOMM Conference on Internet Measurement, pp. 315–320 (2007)
8. Koroniotis, N., Moustafa, N., Sitnikova, E., Turnbull, B.: Towards the development of realistic botnet dataset in the internet of things for network forensic analytics: bot-IOT dataset. Future Gen. Comput. Syst. **100**, 779–796 (2019)
9. Lo, W.W., Layeghy, S., Sarhan, M., Gallagher, M., Portmann, M.: E-graphsage: a graph neural network based intrusion detection system for IOT. In: NOMS 2022–2022 IEEE/IFIP Network Operations and Management Symposium, pp. 1–9. IEEE (2022)
10. Lotfollahi, M., Siavoshani, M.J., Zade, R.S.H., Saberian, M.: Deep packet: a novel approach for encrypted traffic classification using deep learning. Soft Comput. **24**(3), 1999–2012 (2019). https://doi.org/10.1007/s00500-019-04030-2
11. Ma, X., et al.:: A comprehensive survey on graph anomaly detection with deep learning. IEEE Trans. Knowl. Data Eng. (2021)
12. Macas, M., Wu, C., Fuertes, W.: A survey on deep learning for cybersecurity: progress, challenges, and opportunities. Comput. Netw. **212**, 109032 (2022). https://doi.org/10.1016/j.comnet.2022.109032
13. Pacheco, F., Exposito, E., Gineste, M., Baudoin, C., Aguilar, J.: Towards the deployment of machine learning solutions in network traffic classification: a systematic survey. IEEE Commun. Surv. Tutor. **21**(2), 1988–2014 (2019). https://doi.org/10.1109/COMST.2018.2883147
14. Parmisano, A., Garcia, S., Erquiaga, M.J.: A Labeled Dataset with Malicious and Benign IOT Network Traffic. Stratosphere Laboratory, Praha, Czech Republic (2020)
15. The Guardian: DDoS attack that disrupted internet was largest of its kind in history, experts say. https://www.theguardian.com/technology/2016/oct/26/ddos-attack-dyn-mirai-botnet
16. Ullah, I., Mahmoud, Q.H.: A scheme for generating a dataset for anomalous activity detection in IoT networks. In: Goutte, C., Zhu, X. (eds.) Canadian AI 2020. LNCS (LNAI), vol. 12109, pp. 508–520. Springer, Cham (2020). https://doi.org/10.1007/978-3-030-47358-7_52

17. Wang, X., Jin, B., Du, Y., Cui, P., Tan, Y., Yang, Y.: One-class graph neural networks for anomaly detection in attributed networks. Neural Comput. Appl. **33**, 12073–12085 (2021)

18. Xu, Z., Huang, X., Zhao, Y., Dong, Y., Li, J.: Contrastive attributed network anomaly detection with data augmentation. In: Advances in Knowledge Discovery and Data Mining: 26th Pacific-Asia Conference, PAKDD 2022, Chengdu, 16–19 May 2022, Proceedings, Part II, pp. 444–457. Springer, Cham (2022). https://doi.org/10.1007/978-3-031-05936-0_35

19. Zheng, J., Li, D.: Gcn-tc: combining trace graph with statistical features for network traffic classification. In: ICC 2019–2019 IEEE International Conference on Communications (ICC), pp. 1–6. IEEE (2019)

20. Zheng, J., Zeng, Z., Feng, T.: Gcn-eta: high-efficiency encrypted malicious traffic detection. Secur. Commun. Netw. **2022**, 1–11 (2022)

21. Zola, F., Segurola-Gil, L., Bruse, J.L., Galar, M., Orduna-Urrutia, R.: Network traffic analysis through node behaviour classification: a graph-based approach with temporal dissection and data-level preprocessing. Comput. Secur. **115**, 102632 (2022)

Cell Segmentation of *in situ* Transcriptomics Data Using Signed Graph Partitioning

Axel Andersson, Andrea Behanova$^{(\boxtimes)}$, Carolina Wählby, and Filip Malmberg

Centre for Image Analysis, Department of Information Technology and SciLifeLab
BioImage Informatics Facility, Uppsala University, Uppsala, Sweden
`{axel.andersson,andrea.behanova,carolina.wahlby,filip.malmberg}@it.uu.se`

Abstract. The locations of different mRNA molecules can be revealed by multiplexed in situ RNA detection. By assigning detected mRNA molecules to individual cells, it is possible to identify many different cell types in parallel. This in turn enables investigation of the spatial cellular architecture in tissue, which is crucial for furthering our understanding of biological processes and diseases. However, cell typing typically depends on the segmentation of cell nuclei, which is often done based on images of a DNA stain, such as DAPI. Limiting cell definition to a nuclear stain makes it fundamentally difficult to determine accurate cell borders, and thereby also difficult to assign mRNA molecules to the correct cell. As such, we have developed a computational tool that segments cells solely based on the local composition of mRNA molecules. First, a small neural network is trained to compute attractive and repulsive edges between pairs of mRNA molecules. The signed graph is then partitioned by a mutex watershed into components corresponding to different cells. We evaluated our method on two publicly available datasets and compared it against the current state-of-the-art and older baselines. We conclude that combining neural networks with combinatorial optimization is a promising approach for cell segmentation of in situ transcriptomics data. The tool is open-source and publicly available for use at https://github.com/wahlby-lab/IS3G.

Keywords: Cell segmentation · in situ transcriptomics · tissue analysis · mutex watershed

1 Introduction

Over the past years, a large number of techniques for spatially resolved multiplexed in situ transcriptomics (IST) have been developed [1,2,5,6,8,10]. These techniques enable the mapping of hundreds of different mRNA molecules directly within tissue samples, allowing the dissection and analysis of cell type heterogeneity while preserving spatial information. These techniques produce large gigapixel-sized images of tissue sections with millions of different spatial

---

A. Andersson and A. Behanova—Contributed equally.

M. Vento et al. (Eds.): GbRPR 2023, LNCS 14121, pp. 139–148, 2023.
https://doi.org/10.1007/978-3-031-42795-4_13

biomarkers for various mRNA molecules. A single experiment can pinpoint the location of hundreds of different types of mRNA molecules with sub-micrometer resolution. The many different types of targeted mRNA molecules, as well as a large number of detected molecules, make visual exploration and analysis challenging. To alleviate, and compute quantitative statistics, a range of tools, as described below, have been developed for grouping the mRNA molecules into groups, such as cells, cell types, or tissue-level compartments.

1.1   Tools for Analyzing IST Data

The analysis of IST data typically starts by assigning mRNA molecules to localized cells. By examining the composition of molecules within cells, it becomes possible to define and analyze different cell types and their spatial organization [3,11,19,22]. Traditionally, cells are located in a nuclear stained image complementary to the IST experiment using techniques such as Stardist [21], Cellpose [18], or the distance-transform watershed approach. Subsequently, the mRNA molecules are assigned to the detected nuclei based on their proximity in space. However, assigning molecules to cells solely based on their spatial proximity to the nucleus may not be optimal due to irregular cell shapes and the asymmetric distribution of detected mRNAs around the nuclei. Moreover, there are situations where clusters of mRNA molecules belonging to a cell are detected in the IST experiment, but the corresponding nucleus lies outside the imaged region of interest. Additionally, if the quality of the nuclear image is poor and the nuclei are not clearly visible, it further complicates the assignment process. In such cases, cell detection based on nuclear staining is not optimal.

To improve the assignment of molecules to already detected cells, Qian et al [17] created a probabilistic cell typing method, pciSeq, where prior known information regarding the molecular composition of different cell types is utilized when assigning molecules and typing cells. Similarly, Prabhakaran et al. [16] introduced Sparcle, a method where mRNAs are assigned to cells through a relatively simple "assign" and "refine" algorithm. However, both Sparcle and pciSeq require that the location of cells is known beforehand. Petukhov et al. [15] introduced Baysor, an extensive probabilistic model that both detects the location of cells and assigns molecules to the cell. Alternatively, there are methods that overcome assigning molecules to cells by simply ignoring the cells, and instead assigning molecules to spatial bins [12,13,20], or using deep learning to learn more abstract features [4,9,14]. Such methods allow the user to easily identify regions of similar molecular compositions (corresponding to semantic segmentation), but statistics on a per-cell level (requiring instance segmentation) are difficult to compute.

1.2   Contribution

Localizing cells and assigning the right molecules to cells is a crucial task in IST. In this context, we introduce In Situ Sequencing SEGmentation (IS3G), a novel tool that jointly identifies the location of cells and assigns mRNA molecules to them, without the need for prior knowledge of cell location or cell type molecular compositions. IS3G operates solely on local mRNA composition and identifies cells by partitioning a signed graph, making it possible to segment cells without relying on nuclear staining. Our results demonstrate that signed graph partitioning can be used to efficiently segment IST data without seeds for cell location or cell types.

2   Methodology

In brief, IS3G utilizes a small neural network to classify whether two mRNA molecules originate from the same cell or not. The posterior probabilities of the classifier are used to determine the strength of attractive (positive) and repulsive (negative) edges in a signed graph. Using a mutex watershed [23], this graph is then partitioned into components that correspond to individual cells, enabling the accurate assignment of molecules to their respective cells.

2.1   Compositional Features

The neural network used to predict the attractive and repulsive edge strengths is trained on local compositions of mRNA molecules. In this section, we explain how to extract such features. We start by setting the notation. In an IST experiment, the i'th detected mRNA molecule can be described with two attributes: a position $p_i \in \mathbb{R}^D$ and a categorical label $l_i \in \mathbb{L}$, where D is the number of spatial dimensions (usually two or three) and $\mathbb{L}$ is the set of targeted mRNA molecules. The categorical label of a molecule can be represented as a one-hot-encoded vector, i.e., $e_i \in \{0,1\}^{|\mathbb{L}|}$. The local composition of mRNA molecules in a neighborhood around the i'th mRNA can thus be computed by simply counting the labels among the k nearest neighbors,

$$x_i = \sum_{j \in \mathcal{N}(p_i)} e_j, \tag{1}$$

where $\mathcal{N}(p_i)$ refers to the k nearest neighbors to the i'th molecule. The parameter k depends on the molecular density and must be tuned so that the compositional features describe molecular patterns on a cellular scale. We found that setting the value of k to roughly one third of the expected molecular count per cell works well. The compositional features x_i will serve as the input to our model used for computing the strength of our attractive and repulsive edges.

To predict the edge strengths between molecules, we train a small Siamese neural network to classify whether two molecules belong to the same cell or not. The network consists of an encoder and a classifier. First, the compositional features are computed for each of the pairs according to Eq. 1. The pair of features, (x_i, x_j), are respectively encoded by the encoder into latent vectors (z_i, z_j) using a simple neural network with two fully-connected layers with ELU activations. Next, the Euclidean distance between the two latent vectors is computed, $\Delta_{i,j} = \|z_i - z_j\|_2$. Based on the distance between the vectors, the network

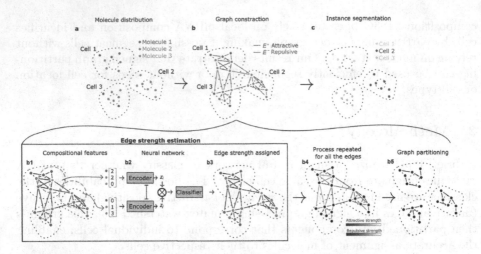

Fig. 1. Explanatory overview of the method. The top row: (**a**) Visualization of the distribution of molecules within three cells, where each color represents a different type of mRNA molecule. (**b**) Attractive edges connect molecules over short distances, whereas repulsive edges connect molecules over greater distances. The weight of the attractive edges indicates the *desire* that the two molecules belong to the same cell, whereas the weight of the repulsive edges indicates a desire that the molecules belong to two different cells. Panel (**b1–b3**) describes how the edge weights are computed. First, compositional vectors are computed for a pair of molecules (**b1**). The pair of compositional vectors, (x_i, x_j), are respectively fed through an encoder, transforming the vectors into latent representations (z_i, z_j) (**b2**). The Euclidean distance between the latent vectors is fed through a classifier, attempting to classify whether the pair of latent vectors are from the same cell or not. The posterior probabilities of the classifier are used to set the strength of the attractive and repulsive edges (**b3–b4**). The signed graph is partitioned using a mutex watershed. The connected components defined by the active attractive edges are labeled individual cell instances (**b5,c**).

attempts to classify whether the sampled molecule pair belongs to the same cell or not. The classifier consisted of a single fully connected layer ending with a Sigmoid function. The posterior probability, y_{ij}, is then used as the attractive and repulsive ($y_{ij} - 1$) edge strengths. However, compositional features (x_i) computed in regions with low concentrations of mRNA molecules are inherently more sensitive to small molecular variations than compositional features computed in high-concentration regions. We, therefore, scale edge weights by the density factor $\rho_{ij} = \min(\|x_i\|_1, \|x_j\|_1)$ to give edges between pairs of mRNA molecules in low concentration regions lower weight than pairs in high concentration regions.

The weights of the encoder and classifier are jointly optimized by minimizing the binary cross entropy using the Adam optimizer [7]. Training data is generated stochastically using the heuristic that two molecules separated by a distance less than R_{cell} are labeled as belonging to the same cell, whereas molecules separated by a distance larger than $2R_{\text{cell}}$ are labeled to belong to two different cells. We trained the network for a maximum of 300 epochs, where one epoch was reached when we sampled the same number of molecule pairs as the total number of molecules in the dataset. However, 300 epochs were never reached since we employed an early stopping procedure. If the loss was non-decreasing for more than 15 epochs, the training would terminate. The model with the lowest loss was saved and used for computing edge strengths. A batch size of 2048 was used for all experiments.

2.2 Graph Construction and Partitioning

In the previous section, we explained how a neural network can be used to compute attractive and repulsive edge strengths. In this section, we will define our set of attractive and repulsive edges. We define our set of attractive edges, E^+, by assuming that molecules separated by a short distance are likely to belong to the same cell. Specifically, E^+ is created by connecting each molecule to all of its five nearest neighbors.

Next, we define our set of repulsive edges between molecules separated by a distance larger than $2R_{\text{cell}}$ and less than $6R_{\text{cell}}$. To save memory, each molecule is only connected with a repulsive edge to 15 randomly selected neighbors within the interval. The repulsive edge weights are set to $-\infty$ for edges between molecules separated by a distance larger than $4R_{\text{cell}}$.

Finally, we use the procedure from the previous section to compute the strength of the edges, leaving us with a graph with signed edge weights. The graph is finally partitioned using a mutex watershed into components corresponding to different cells.

The methodology is shown as an illustrative example in Fig. 1.

2.3 Pre and Post Processing

To speed up the segmentation we first remove markers in low-density regions, as these markers are likely extracellular. We identify these markers automatically by computing the distance to the 15'th nearest neighbor. The distance is then clustered using a Gaussian mixture model. Markers belonging to the component with the larger mean are deemed extracellular and removed. After the segmentation, we discard cells with fewer than n_{min} number of molecules.

2.4 Visualization

The segmented cells can be visualized in two ways: by assigning a color to each marker based on the cell it belongs to, or by outlining the segmented cell with a contour. To generate the contours, we employ an algorithm similar to the one described in [15]. For each cell, we calculate the marker frequency within small spatial bins placed on a regular grid. This process is repeated for all detected cells, resulting in a multi-channel count image. Each pixel in the image represents the marker count in a bin for a specific cell. The count image is spatially smoothed using a Gaussian filter and then transformed into a 2D labeled mask by identifying the index of the maximum value across the channels. Each label in the mask thus corresponds to detected cells. Next, we find the contours of each cell in the labeled mask. The contour pixels for each cell are finally filtered by taking the longest path of the minimum spanning tree connecting the points.

3   Experiments

We perform experiments on two publicly available datasets: An osmFISH [1] and an In Situ Sequencing (ISS) [17] dataset and compare with authors' original segmentation as well as Baysor [15]—the current state-of-the-art.

3.1   osmFISH

We first studied the osmFISH dataset [1]. This dataset consists of around two million molecules of 35 different types. Also included in this dataset is a segmentation produced by the original authors [1], here referred to as the Codeluppi method. We ran IS3G using $R_{cell} = 8$ mm and $k = 35$. We also ran Baysor [15] on the dataset, using the parameters provided in their osmFISH example. Baysor can be seeded with prior information regarding the nuclei segmentation (as described in [15]), as such, we run Baysor both with and without such a prior. For each cell segmentation method, we filter out cells containing fewer than $n_{min} = 30$ molecules. First, we look at the number of cells detected by each of the methods as well as the fraction of assigned molecules. This is shown in Fig. 2a and Fig. 2b respectively. As seen, IS3G finds roughly the same number of cells as Baysor and Baysor with prior, but significantly more than the original publication [1].

Next, we wanted to investigate if we find cells in the same location as the other methods. To do this we first matched our detected cells with the other methods' detected cells based on Sørensen-Dice index (Dice index). If two cells identified by two methods contain exactly the same molecules, the Dice index is one, and zero if no molecules overlap between the two cells. We match the cells between the two methods by maximizing the average Dice score across all detected cells using the Hungarian algorithm. Figure 2c shows the distribution of Dice indices between matched cells detected using IS3G versus Baysor, Baysor with prior and Codeluppi et al. [1]. The median Dice index between IS3G detected cells matched with Baysor detected cells was 0.8. Finally, Fig. 2d shows some examples of the segmentation done by the different methods. The full dataset with segmentation results from all the mentioned techniques can be found here: https://tissuumaps. scilifelab.se/osmFISH.html

3.2   In Situ Sequencing

Secondly, we studied the dataset by Qian et al [17]. This dataset comprises around 1.4 million detected molecules of 84 different types. As for post-processing, we filter out cells containing fewer than $n_{min} = 8$ molecules. We used an $R_{cell} = 10$ μm and $k = 8$. Figure 3a shows the number of cells detected

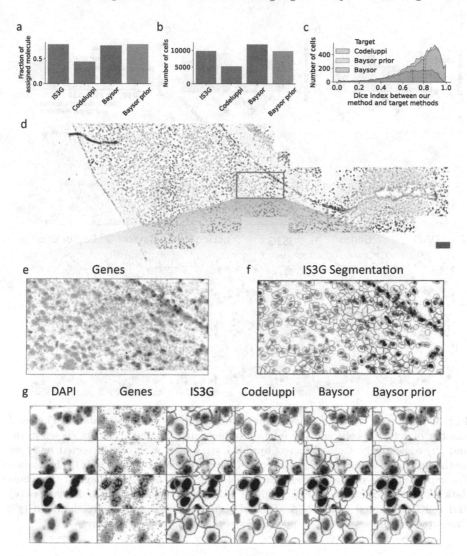

Fig. 2. Results of various segmentation techniques applied on osmFISH dataset. The total number of detected cells and the fraction of molecules assigned cells are shown in **a** and **b** respectively. Cells detected by IS3G are paired with cells detected by other methods. The distribution of Dice indices of the paired cells is shown in **c**. The dashed line represents the median Dice index. Panel **d** shows a zoomed out view of the DAPI image (100 μm scale bar), with zoom-ins showing the distribution of gene markers (**e**) and IS3G segmented cells (**f**). A series of segmentation examples are shown in **d**. Presented techniques are IS3G, Codeluppi, Baysor, and Baysor with DAPI.

by IS3G, Baysor, Baysor with prior, and pciSeq. We note that IS3G finds roughly the same number of cells as Baysor and Baysor prior, and significantly more than pciSeq.

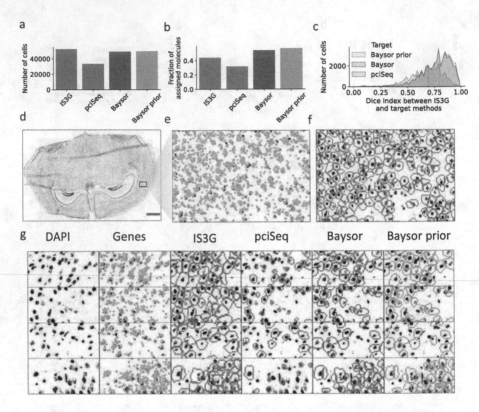

Fig. 3. Results of various segmentation techniques applied on ISS dataset. The total number of detected cells and the fraction of molecules assigned cells are shown in (**a**) and (**b**) respectively. Cells detected by IS3G are paired with cells detected by other methods. The distribution of Dice indices of the paired cells is shown in (**c**). The dashed line represents the median Dice index. Panel (**d**) shows the DAPI image of the whole data set (1 mm scale bar) with highlighted zoom-in sections showing gene markers (**e**) and IS3G segmented cells (**f**). A series of segmentation examples are shown in (**g**). Presented techniques are IS3G, pciSeq, Baysor, and Baysor with prior.

Figure 3c shows the distribution of Dice indices when matching molecules assigned to our segmented cells with the other methods. The dashed lines indicate the median. Finally, Fig. 3d shows some examples of the segmentation done by the different methods. The full dataset with segmentation results from all the mentioned techniques can be found here: https://tissuumaps.scilifelab.se/ISS.html

4   Discussion

We have presented a simple technique for segmenting cells in IST data. IS3G differs from other approaches that need prior cell segmentation, seeds, or pre-determined cell types. Instead, IS3G directly extracts features from the data by utilizing a simple neural network. We tested IS3G on two datasets, and it achieved performance comparable to the current state of the art, see Fig. 2c and Fig. 2c, showing that signed graph partitioning can be used to efficiently segment cells in IST data.

The deep learning model used to predict edge weights is very basic and likely not optimal. It was trained on "already aggregated data," which refers to compositional features obtained by computing the weighted frequency of molecules in circular neighborhoods, see Eq. 1. A graph neural network may be more suitable for this application since it can also learn the weights used in the aggregation.

IS3G requires that the user provides a rough estimate of the cell radius, i.e., R_{cell}. This parameter governs the bandwidth of the Gaussian kernel used when computing the compositional features. However, here we have assumed that the size of each cell is approximately the same. In practice, we have noticed that the size of the cells, or more precisely, the size of the mRNA point-cloud surrounding the cells, can vary between cells. Potentially, the segmentation could be improved by considering an adaptive bandwidth or using a graph-neural network that can extract features across multiple scales.

While not used explicitly herein, the mutex watershed algorithm provides a convenient way to specify mutually exclusive constraints between specific types of markers. This could be particularly beneficial in regions where it is difficult to differentiate between cells based solely on mRNA composition but a clear distinction can be made based on their nuclei. In such scenarios, IS3G may identify cells with multiple nuclei. However, if the user has supplementary markers indicating the position of nuclei, infinitely repulsive edges can be incorporated between these markers to explicitly ensures that each cell contains only one nucleus.

References

1. Codeluppi, S., et al.: Spatial organization of the somatosensory cortex revealed by osmFISH. Nat. Meth. **15**(11), 932–935 (2018). https://doi.org/10.1038/s41592-018-0175-z
2. Eng, C.H.L., et al.: Transcriptome-scale super-resolved imaging in tissues by RNA seqFISH. Nature **568**(7751), 235–239 (2019). https://doi.org/10.1038/s41586-019-1049-y
3. Hao, Y., et al.: Integrated analysis of multimodal single-cell data. Cell **184**(13), 3573–3587 (2021)
4. Hu, J., et al.: SpaGCN: integrating gene expression, spatial location and histology to identify spatial domains and spatially variable genes by graph convolutional network. Nat. Meth. **18**(11), 1342–1351 (2021)
5. Janesick, A., et al.: High resolution mapping of the breast cancer tumor microenvironment using integrated single cell, spatial and in situ analysis of FFPE tissue (2022). https://doi.org/10.1101/2022.10.06.510405
6. Ke, R., et al.: In situ sequencing for RNA analysis in preserved tissue and cells. Nat. Meth. **10**(9), 857–860 (2013). https://doi.org/10.1038/nmeth.2563
7. Kingma, D.P., Ba, J.: Adam: a method for stochastic optimization (2014). https://doi.org/10.48550/ARXIV.1412.6980

8. Lee, H., Salas, S.M., Gyllborg, D., Nilsson, M.: Direct RNA targeted in situ sequencing for transcriptomic profiling in tissue. Sci. Rep. **12**(1), 7976 (2022). https://doi.org/10.1038/s41598-022-11534-9
9. Li, J., Chen, S., Pan, X., Yuan, Y., bin Shen, H.: CCST: cell clustering for spatial transcriptomics data with graph neural network (2021)
10. Moffitt, J.R., et al.: Molecular, spatial, and functional single-cell profiling of the hypothalamic preoptic region. Science **362**(6416), eaau5324 (2018). https://doi.org/10.1126/science.aau5324
11. Palla, G., et al.: Squidpy: a scalable framework for spatial omics analysis. Nat. Meth. **19**(2), 171–178 (2022)
12. Park, J., et al.: Cell segmentation-free inference of cell types from in situ transcriptomics data. Nat. Commun. **12**(1), 4103 (2021). https://doi.org/10.1038/s41467-021-23807-4
13. Partel, G., et al.: Automated identification of the mouse brain's spatial compartments from in situ sequencing data. BMC Biol. **18**(1), 1–14 (2020)
14. Partel, G., Wählby, C.: Spage2vec: unsupervised representation of localized spatial gene expression signatures. FEBS J. **288**(6), 1859–1870 (2020). https://doi.org/10.1111/febs.15572
15. Petukhov, V., et al.: Cell segmentation in imaging-based spatial transcriptomics. Nat. Biotechnol. **40**(3), 345–354 (2022)
16. Prabhakaran, S.: Sparcle: assigning transcripts to cells in multiplexed images. Bioinf. Adv. **2**(1), vbac048 (2022)
17. Qian, X., et al.: Probabilistic cell typing enables fine mapping of closely related cell types In situ. Nat. Meth. **17**(1), 101–106 (2019). https://doi.org/10.1038/s41592-019-0631-4
18. Stringer, C., Wang, T., Michaelos, M., Pachitariu, M.: Cellpose: a generalist algorithm for cellular segmentation. Nat. Meth. **18**(1), 100–106 (2020). https://doi.org/10.1038/s41592-020-01018-x
19. Teng, H., Yuan, Y., Bar-Joseph, Z.: Clustering spatial transcriptomics data. Bioinformatics **38**(4), 997–1004 (2022)
20. Tiesmeyer, S., Sahay, S., Müller-Bötticher, N., Eils, R., Mackowiak, S.D., Ishaque, N.: SSAM-lite: a light-weight web app for rapid analysis of spatially resolved transcriptomics data. Front. Genet. **13**, 785877 (2022). https://doi.org/10.3389/fgene.2022.785877
21. Weigert, M., Schmidt, U., Haase, R., Sugawara, K., Myers, G.: Star-convex polyhedra for 3D object detection and segmentation in microscopy. In: The IEEE Winter Conference on Applications of Computer Vision (WACV), March 2020 (2020). https://doi.org/10.1109/WACV45572.2020.9093435
22. Wolf, F.A., Angerer, P., Theis, F.J.: SCANPY: large-scale single-cell gene expression data analysis. Genome Biol. **19**(1), 1–5 (2018)
23. Wolf, S., et al.: The mutex watershed and its objective: efficient, parameter-free graph partitioning. IEEE Trans. Pattern Anal. Mach. Intell. **43**(10), 3724–3738 (2020)

Graph-Based Representation for Multi-image Super-Resolution

Tomasz Tarasiewicz(✉) 🆔 and Michal Kawulok 🆔

Department of Algorithmics and Software, Silesian University of Technology,
Gliwice, Poland
{tomasz.tarasiewicz,michal.kawulok}@polsl.pl

Abstract. Multi-image super-resolution is a challenging computer vision problem that aims at recovering a high-resolution image from its multiple low-resolution counterparts. In recent years, deep learning-based approaches have shown promising results, however, they often lack the flexibility of modeling complex relations between pixels, permutability of the input data, or they were designed to process a specific number of input images. In this paper, we propose an improved version of our earlier graph neural network that benefits from permutation-invariant graph-based representation of multiple low-resolution images. Importantly, we demonstrate that our solution allows for performing reconstruction from a set of heterogeneous input images, which is not straightforward for other state-of-the-art techniques. Such flexibility is a crucial feature for practical applications, which is confirmed qualitatively and quantitatively for a set of real-world (rather than simulated) input images.

Keywords: Multi-image super-resolution · Image fusion · Graph neural network

1 Introduction

Super-resolution (SR) reconstruction is a common term for a variety of techniques whose common goal is to generate a high-resolution (HR) image from a low-resolution (LR) observation. The latter may take the form of a single image or multiple images presenting the same scene. Single-image SR (SISR) techniques are easy to apply, as they do not require multiple images to operate, but they are severely ill-posed, since in most cases an LR image can be super-resolved into a variety of HR images. Multi-image SR (MISR) relies on the assumption that every LR image carries a different portion of HR information. Therefore, by means of information fusion, an HR image can be reconstructed more reliably than with SISR.

This research was supported by the National Science Centre, Poland, under Research Grant No. 2019/35/B/ST6/03006. MK was supported by the Silesian University of Technology, Poland funds through the Rector's Research and Development Grants No. 02/080/RGJ22/0024. TT benefits from the European Union scholarship through the European Social Fund (grant POWR.03.05.00-00-Z305).

M. Vento et al. (Eds.): GbRPR 2023, LNCS 14121, pp. 149–159, 2023.
https://doi.org/10.1007/978-3-031-42795-4_14

1.1   Related Work

The best-performing SISR and MISR methods are underpinned with convolutional neural networks (CNNs) that learn the reconstruction procedure from the matched LR and HR images showing the same scene. The use of CNNs for SISR has been widely explored since 2014 when Dong et al. demonstrated that the processing pipeline of sparse-coding-based SR [17] can be viewed as a deep CNN. However, applying CNNs for MISR is not that straightforward, mainly due to the problems with appropriate input data representation. In 2019, we introduced the EvoNet framework [7], which preprocesses the input LR images using CNN-based SISR techniques, prior to evolutionary multi-image fusion [6]. Such preprocessing enhances the performance of MISR, but this approach has been outperformed with end-to-end deep learning solutions, proposed to address the PROBA-V SR Challenge that European Space Agency (ESA) organized in 2018–19 [8], and published the first large-scale real-world dataset for MISR. The DeepSUM network [9], later enhanced with non-local operations [10], was the first end-to-end MISR network. It is composed of three parts that (i) extract deep features from each LR image, (ii) co-register the feature maps with sub-pixel precision, and (iii) perform their final fusion, whose outcome is added to the image obtained as an average of bicubically-upsampled LR inputs. DeepSUM assumes a fixed number of LR inputs and requires long training, being the result of fusing the upsampled LR images. These downsides were addressed in other MISR solutions, including HighRes-Net, which combines the latent LR representations in a recursive manner to obtain the global representation, which is upsampled to obtain the super-resolved image [2]. Also, the attention mechanism was found useful for selecting the most valuable features extracted from LR inputs in the residual attention multi-image SR (RAMS) network [12], and a recurrent network with gated recurrent units was proposed in [11]. An et al. focused on simplifying the training with the use of transformers [1], and reported competitive results obtained with their TR-MISR network for the PROBA-V dataset. Recently, Valsesia and Magli showed with their PIUnet [16] that enforcing permutation invariance within a set of LR inputs leads to significant improvements in super-resolving multiple images.

In [15], we proposed to represent an input set of LR images as a graph which is processed with a graph neural network (GNN) that produces the super-resolved image. It leverages the inherent structure of the LR images and allows for flexible connectivity between any pair of pixels on the graph. However, the graph is transformed to a rectangular form prior to the upsampling performed with the pixel shuffle operation, and this process assumes uniform pixel density across the image (locally, the graph must inherit identical structure around each pixel). This means that such a network can benefit from sub-pixel shifts in the spatial domain, but it is assumed that the input images are of an identical size and that they are not rotated between each other. Importantly, the same limitation is imposed by other state-of-the-art techniques.

1.2 Contribution

In this work, we address the important limitations of the technique proposed in our earlier work [15] and we study the practical benefits of graph-based representation for MISR. In particular, our contribution is threefold:

1. We propose Magnet++ which is an improved version of Magnet-our earlier GNN for MISR. The primary improvement in Magnet++ lies in its upsampling which leverages the graph structure of the input data for more efficient and accurate performance. Unlike our previous model, where the upsampling operation was performed after the graph had been transformed to a rectangular form and then upsampled using the pixel shuffle algorithm, Magnet++ utilizes the inherent graph structure for a more effective and precise upsampling process.
2. We report the results obtained for a real-world dataset with scenes composed of multiple original low-resolution images captured by the PROBA-V satellite showing the same area of Earth, coupled with a high-resolution image acquired by a different sensor installed on the same satellite.
3. We argue that in Magnet++, the proposed graph-based representation coupled with improved upsampling allows for performing super-resolution from a set of heterogeneous low-resolution images, and we report the initial, yet encouraging, results of our experimental study that confirm such capability.

Overall, the proposed technique (Sect. 2) provides a new perspective on SR by exploiting the graph representation of LR images. The results of our experiments (Sect. 3) confirmed that it allows for running the reconstruction from a set of heterogeneous images of varying sizes without the need of resampling them to a uniform regular grid, which has important practical implications. We will make Magnet++ implementation publicly available upon paper acceptance.

2 Proposed Method

This section covers various techniques for converting a stack of LR images into a single graph (Sect. 2.1), as well as a simple GNN designed to process this type of data (Sect. 2.2). Additionally, we examine key features of both the GNN and the proposed data representation.

2.1 Data Representation

Multiple images of the same scene may not be identical even when taken almost immediately one after another due to various factors such as sensor noise, changes in lighting conditions, camera shake or movement, and lens distortion. These factors can introduce variations in the captured images, resulting in differences in pixel values and overall image appearance. In MISR, this is an especially useful phenomenon as it allows for the extraction and fusion of complementary information contained in each image. One of the most common differences between such images is their displacement with respect to the point of reference. These displacements can span over multiple pixels or have values lower than one unit (subpixel shifts). To address this issue, most of the current state-of-the-art MISR methods apply different image co-registration techniques [5,9] in order to modify the input images so that each pixel of a single image represents exactly the same point on a 2D plane as its counterparts from other observations. These data modifications may lead to a loss of important information, which can hinder successful scene reconstruction. In our approach, however, we do not perform any modifications on the input data; hence the initial information is preserved in its entirety.

Node Positioning. To create a single graph representation of a stack of N input LRs, we need to correctly place them on a 2D plane. To accomplish this, we compute the displacement vectors for each LR with respect to a randomly chosen image from the stack. This can be achieved using readily available registration algorithms or a trained neural network. We then recenter the displacement vectors by subtracting the total mean of the displacements from each of them, thereby minimizing any bias towards the reference image that has no initial shifts. Next, we place the images on a 2D plane by assigning each pixel a discrete position (x, y), where x and y correspond to indices of a pixel in an input image LR_n, assuming $LR_n \in \mathbb{R}^{H \times W}$ and $n \in (1, 2, ..., N)$. At this stage, each pixel shares its discrete position with corresponding pixels from other images, so we adjust their location by subtracting a displacement vector assigned for each LR.

It is worth noting that the proposed method is designed for applications where each image has the same shape and resolution. However, we can enhance this approach by incorporating more advanced techniques for assigning node positions, such as encoding geospatial coordinates for remote sensing applications. This would enable us to create graphs using LR images of varying resolutions or sizes rather than being restricted to a constant shape for each LR image or even combining images rotated with respect to each other.

Creating Connections. In the context of graph theory, an edge denotes a linkage or a bond connecting a pair of vertices or nodes in a graph. The connection is represented as a line or a curve, serving to indicate the relationship between the vertices. Depending on whether it signifies a one-way or a two-way connection between the vertices, an edge can be either directed or undirected. Concerning GNNs, the presence of an edge signifies that two nodes are connected, and the edge's direction specifies whether the node is a source of information or a receiver. Furthermore, an edge can be classified based on the relation it represents, given a weight to express its strength, or endowed with a set of features.

In the context of the MISR problem, a typical approach would be to establish edges solely between neighbouring nodes, given the relatively higher importance of local information over global one. In a 2D tensor representation of an image, each pixel has eight immediate neighbours, except those situated at the image's boundary. However, in the case where all nodes across multiple images are placed on a single, continuous Euclidean plane, determining their neighbourhood relationships becomes non-trivial. One potential solution, which we embraced in this work, is to link nodes based on their Euclidean distance, with only those pairs of nodes that fall within a pre-specified radius r being connected. For our experiments, we set $r = 1$. This ensures a thoughtful equilibrium between maintaining computational manageability and preserving the local integrity of the image by connecting each pixel to its direct neighbors. It's important to note that any increase in the radius would escalate the connections, potentially overcomplicating the graph and intensifying computational demands.

Message Passing. In geometric deep learning, message passing is the fundamental operation used to update node states by propagating information. The form of the message-passing function can vary depending on the architecture of the GNN. However, the essential idea is to utilize the available graph information, such as node and edge features, to update node states in a way that reflects the graph's structure and relationships. In our research, we have discovered that the spline-based convolutional operator [4] is especially effective when integrated into the MISR GNN. It works similarly to the sliding-window approach of standard convolutional layers, however, if nodes are placed on a continuous plane, applying a $K \times K$ convolutional kernel with only K^2 probing points would be impossible. The spline-based convolution overcomes this issue by using B-spline basis functions to create a continuous kernel with K^2 trainable parameters and interpolate their value depending on the position of a node being probed. Additionally, the span of a standard convolution is closely related to the size of a kernel and its dilation, whereas the spline-based operator allows for an operational distance independent of the kernel size. This means that the larger the kernel, the denser it samples node values assuming a constant operational distance.

2.2 Graph Neural Network

The proposed architecture, Magnet++ is inspired by Magnet[15], which leverages spline-based convolutions and a message-passing GNN approach to capture spatial information in a graph and node features. Similar to Magnet, our architecture features a feature extraction layer, followed by a shrinking layer to reduce trainable parameters and a convolutional block with skip connections, all of which utilize spline-based convolutions. However, it diverges in the upscaling method, which more effectively capitalizes on the unique characteristics of our data representation.

In Magnet, the upscaling was performed by firstly max-pooling nodes that correspond to the same pixel position of the LR images in the $N \times W \times H$ tensor form. This operation reduced the number of nodes N times. The resulting nodes were then assumed to be positioned in a grid-like manner and transformed into a $W \times H \times f$ tensor, where f represents a number of features produced by the spline convolutional block. It was later processed by a 2D convolutional block and passed to the pixel shuffle layer, which is responsible for rearranging the tensor's features to increase the spatial resolution of the image.

In contrast, Magnet++ leverages a combination of graph theory and spline-based convolutions to perform image upscaling. Firstly, it takes a graph, combined with multiple LR images, which is processed by the formerly mentioned feature extraction and shrinking layers followed by a convolutional residual block. Then, on top of that, we overlay a new regular grid-like graph consisting of S^2WH nodes where S stands for an upscaling factor of the network. The new nodes are initialized with zeros and connected to the original nodes lying within a specified radius $r = \sqrt{2}$. Every such connection is created in a one-way manner, meaning that each already existing node is the source of information and the overlayed nodes are the targets. This creates a single bipartite graph consisting of two sets of nodes, where there is no existing connection between any pair of nodes belonging to the same set. We then apply a single spline-based convolutional layer activated by the ReLU function to collect the information from the source nodes. Finally, we separate the overlayed grid-like graph from the other nodes and transform it into a single tensor of shape $SW \times SH$. Thanks to this approach, we ensure that the upscaling method does not introduce any loss of information contained in the input graph, which is not guaranteed in the Magnet architecture due to the use of the max-pooling algorithm.

Since the bipartite upscaling operation returns a single rectangular tensor, it makes it suitable to apply the standard 2D convolutional layers instead of the spline-based ones. Thus, we further process the data using a single residually-connected convolutional block followed by a convolutional layer with a single kernel producing the final super-resolved image. The architecture diagram is presented in Fig. 1.

3   Experiments

In this study, we utilized the renowned PROBA-V dataset [8] for both the training and evaluation of our model. The PROBA-V dataset, with its rich collection of satellite data from different regions worldwide, has been instrumental in shaping the evolution of many contemporary MISR techniques. The dataset was carefully curated for a MISR challenge, focusing on the super-resolution of low-resolution images by a factor of $3\times$. Each scene in the dataset is representative of multiple distinct observations. Our study particularly focused on data pertaining to the RED spectral band. Despite the provision of separate training and testing subsets in the dataset, we faced a limitation in the lack of high-resolution images in the test subset, restricting our ability for local model evaluation. As a solution to this challenge, we opted to randomly divide the training subset, which included numerous scenes, into three separate sections for training, validation, and testing in a ratio of 80:10:10.

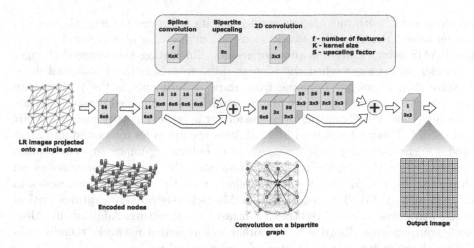

Fig. 1. Architecture of the proposed Magnet++ graph neural network. Multiple input images are co-registered and aligned into a graph that is processed with a graph neural network.

We evaluated our model against two state-of-the-art networks that achieved high rankings on the PROBA-V challenge—HighRes-Net [2] and RAMS [12]. To ensure a fair comparison, we used each model's exact code available in the official online repositories. However, for RAMS, we had to convert the code from Tensorflow implementation to PyTorch in order to incorporate it into our environment. We conducted training of each of the previously mentioned models using our own dataset splits and the same hyperparameters as reported in their original publications rather than utilizing their trained weights published in official repositories. We embraced this approach as the training and validation subsets used in the original publications differed from ours, resulting in different permutations of the available data. Directly using the trained weights could introduce biases towards specific examples and adversely affect performance. By conducting training from the beginning, we ensured a fair and unbiased comparison between our proposed model and the state-of-the-art methods. Magnet++ was trained over 100 epochs using a batch size of 24 and a learning rate of $5 \cdot 10^{-4}$. The loss function was defined as the corrected (in terms of shifts in the brightness and spatial domains) peak signal-to-noise ratio (cPSNR), which is the same metric employed for the official evaluation of models submitted in the PROBA-V challenge. Also, the same corrections are applied to the structural similarity index (SSIM) and learned perceptual image patch similarity (LPIPS) [18], leading to the cSSIM and cLPIPS metrics. The experiments were conducted in Python 3.9.12 using the PyTorch 1.11 and PyTorch Geometric 2.1.0 [3] libraries and were executed on the NVIDIA RTX 3090Ti GPU with 24 GB VRAM.

The results of our experiments on the PROBA-V dataset, using nine LR observations as input, are presented in Table 1. Interestingly, these results contrast with our previous findings on simulated data as detailed in [15]. In our previous work, both our Magnet model and its improved version, Magnet++, outperformed the RAMS network. However, in the case of real-world images, the RAMS network delivers superior results, illustrating the nuanced changes in model performance when the context shifts from simulated to actual data. Despite this, a notable outcome from these experiments is the performance increase when we move from the Magnet model to its enhanced version. The latter model substantially outperforms the former in critical metrics such as cPSNR and cSSIM. This performance boost validates the upgrades integrated into Magnet++, demonstrating their effectiveness in enhancing image super-resolution quality. While the RAMS model demonstrates the highest scores across all three metrics, our Magnet++closely follows with the second-highest scores in cPSNR and cSSIM. This suggests that Magnet++effectively captures critical image attributes such as brightness, contrast, and structure. Additionally, Magnet++outperforms HighRes-Net, another well-regarded network, thereby validating the robustness and potential of our improved model.

Table 1. The scores obtained for the original PROBA-V dataset with nine LR images in each scene. The best scores are boldfaced, and the second-best—underlined.

Model	cPSNR	cSSIM	cLPIPS
Bicubic	35.42	0.898	0.313
HighRes-Net [2]	37.39	0.928	<u>0.161</u>
RAMS [12]	**38.49**	**0.941**	**0.160**
Magnet [15]	37.52	0.930	0.188
Magnet++	<u>38.16</u>	<u>0.936</u>	0.179

In order to verify the flexibility of the graph-based representation, we conducted an experiment focused on processing input images of different sizes. To accomplish this, we selected three low-resolution images from each scene having the highest clear-pixel ratio. For two of the images, we created four subsampled versions of each, with both dimensions reduced to half the original size. This was accomplished by sampling every second row and column, using four unique initial unitary displacements, ensuring that each pixel from the original image is retained in one of the resulting images without any duplicates. Consequently, we obtained nine distinct and heterogeneous LR images (in terms of their dimension), with one remaining unaltered and eight reduced to a quarter of their original size (in terms of the number of pixels). Importantly, this operation did not result in any loss of information contained in the original three low-resolution images, as it merely redistributed it between separate images.

For Magnet++, it was essential to encode each pixel's position within the 2D space accurately. By doing so, the resulting input graph effectively replicates the scenario of passing three original LR images to the network, maintaining the desired graph structure. This task was readily accomplished through the computation of displacement vectors between each pair of corresponding original LR

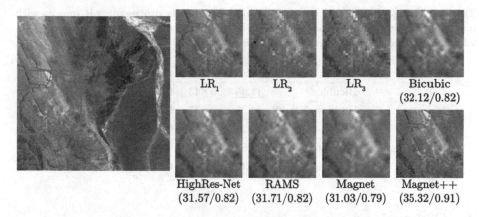

Fig. 2. Visual comparison of RED imgset0337 scene with LR images of varying dimensions. The values in brackets correspond to cPSNR and cSSIM metrics, respectively. LR_1 is the original LR image, while LR_2 and LR_3 are obtained by decomposing another LR image into four counterparts.

images, incorporating the initial displacement applied during the subsampling process. It is important to emphasize that this specific case of encoding node positions for images of varying dimensions benefits from the fact that the decomposed LR images are derived from the original LR image, but this was intended to provide a proof of concept demonstrating the Magnet++ capabilities of processing a stack of heterogeneous images.

Contrary to Magnet++, alternative networks (including our earlier Magnet) enforce restrictions on input data, requiring uniform dimensions for all LR images. To overcome this limitation, we resampled each smaller LR image to the size of the single original LR image in the stack using nearest neighbor interpolation to prevent any loss of information. Subsequently, we fed this stack of uniformly sized images to the other networks. In Table 2, we report the obtained quantitative results. Lower scores (compared with those in Table 1) result from using information from three images instead of nine, and because of the fact that the information was further spread across nine non-uniform LR images. However, it can be seen that Magnet++ retrieved substantially better scores than other models (for all the metrics), as it effectively utilized the information in the stack of LR images in a manner that other models could not achieve due to their limitations. The qualitative results, an example of which is presented in Fig. 2, confirm that Magnet++ leads to much better outcome, showing definitely more details in the reconstructed image.

Table 2. Image similarity metrics computed for models processing LR images with non-uniform dimensions as input data. The highest scores are highlighted in bold, while the second-highest scores are underlined.

Model	cPSNR	cSSIM	cLPIPS
Bicubic	<u>34.35</u>	0.833	0.560
HighRes-Net [2]	34.27	<u>0.837</u>	<u>0.436</u>
RAMS [12]	34.21	0.835	0.506
Magnet [15]	33.62	0.819	0.489
Magnet++	**35.88**	**0.891**	**0.256**

4 Conclusions

In this paper, we proposed an enhanced version of our graph neural network for multi-image super-resolution. Our tests over a set of real-world images confirmed that it achieves competitive quantitative and qualitative results and can operate from a set of heterogeneous low-resolution images. Although in the study reported here, such heterogeneity was limited to the varying image size, the input graph can be assembled from any (even non-regular) images without the need for their resampling to a regular grid [13]. Therefore, our ongoing work is focused on verifying the benefits of graph-based representation on an experimental basis and on introducing further improvements to our technique, including the attention modules which may help select the most useful nodes in each local neighborhood. We believe that our approach, which is based on graph theory and deep learning, can be adapted for solving other image reconstruction problems, including SR of multispectral images with varying resolution of spectral bands [14].

References

1. An, T., Zhang, X., Huo, C., Xue, B., Wang, L., Pan, C.: TR-MISR: multiimage super-resolution based on feature fusion with transformers. IEEE J-STARS **15**, 1373–1388 (2022)
2. Deudon, M., Kalaitzis, A., et al.: HighResnet: recursive fusion for multi-frame super-resolution of satellite imagery. arXiv preprint arXiv:2002.06460 (2020)
3. Fey, M., Lenssen, J.E.: Fast graph representation learning with PyTorch Geometric. In: ICLR Workshop on Representation Learning on Graphs and Manifolds (2019)
4. Fey, M., Lenssen, J.E., Weichert, F., Müller, H.: SplineCNN: fast geometric deep learning with continuous B-spline kernels. In: Proceedings of the IEEE CVPR, pp. 869–877 (2018)
5. Guizar-Sicairos, M., Thurman, S.T., Fienup, J.R.: Efficient subpixel image registration algorithms. Opt. Lett. **33**(2), 156–158 (2008)
6. Kawulok, M., Benecki, P., Kostrzewa, D., Skonieczny, L.: Evolving imaging model for super-resolution reconstruction. In: Proceedings of the GECCO, pp. 284–285. ACM, New York (2018)

7. Kawulok, M., Benecki, P., Piechaczek, S., Hrynczenko, K., Kostrzewa, D., Nalepa, J.: Deep learning for multiple-image super-resolution. IEEE GRSL 17(6), 1062–1066 (2020)
8. Märtens, M., Izzo, D., Krzic, A., Cox, D.: Super-resolution of PROBA-V images using convolutional neural networks. Astrodynamics 3(4), 387–402 (2019)
9. Molini, A.B., Valsesia, D., Fracastoro, G., Magli, E.: DeepSUM: deep neural network for super-resolution of unregistered multitemporal images. IEEE TGRS 58(5), 3644–3656 (2020)
10. Molini, A.B., Valsesia, D., Fracastoro, G., Magli, E.: DeepSUM++: non-local deep neural network for super-resolution of unregistered multitemporal images. In: Proceedings of the IEEE IGARSS, pp. 609–612 (2020)
11. Rifat Arefin, M., et al.: Multi-image super-resolution for remote sensing using deep recurrent networks. In: Proceedings of the IEEE CVPR Workshops, pp. 206–207 (2020)
12. Salvetti, F., Mazzia, V., Khaliq, A., Chiaberge, M.: Multi-image super resolution of remotely sensed images using residual attention deep neural networks. Remote Sens. 12(14), 2207 (2020)
13. Seiler, J., Jonscher, M., Schöberl, M., Kaup, A.: Resampling images to a regular grid from a non-regular subset of pixel positions using frequency selective reconstruction. IEEE Trans. Image Process. 24(11), 4540–4555 (2015)
14. Tarasiewicz, T., et al.: Multitemporal and multispectral data fusion for super-resolution of Sentinel-2 images. arXiv preprint arXiv:2301.11154 (2023)
15. Tarasiewicz, T., Nalepa, J., Kawulok, M.: A graph neural network for multiple-image super-resolution. In: Proceedings of the IEEE ICIP, pp. 1824–1828 (2021)
16. Valsesia, D., Magli, E.: Permutation invariance and uncertainty in multitemporal image super-resolution. IEEE TGRS 60, 1–12 (2022)
17. Yang, J., Wright, J., Huang, T.S., Ma, Y.: Image super-resolution via sparse representation. IEEE TIP 19(11), 2861–2873 (2010)
18. Zhang, R., Isola, P., Efros, A.A., Shechtman, E., Wang, O.: The unreasonable effectiveness of deep features as a perceptual metric. In: Proceedings of the IEEE/CVF CVPR (2018)

Reducing the Computational Complexity of the Eccentricity Transform of a Tree

Majid Banaeyan⬛ and Walter G. Kropatsch(✉)⬛

Pattern Recognition and Image Processing Group, TU Wien, Vienna, Austria
{majid,krw}@prip.tuwien.ac.at

Abstract. This paper proposes a novel approach to reduce the computational complexity of the eccentricity transform (ECC) for graph-based representation and analysis of shapes. The ECC assigns to each point within a shape its geodesic distance to the furthest point, providing essential information about the shape's geometry, connectivity, and topology. Although the ECC has proven valuable in numerous applications, its computation using traditional methods involves heavy computational complexity. To overcome this limitation, we present a method that computes the ECC of a tree, significantly reducing the computational complexity from $\mathcal{O}(n^2 log(n))$ to $\mathcal{O}(b)$, where n and b are the numbers of vertices and branching points in the tree, respectively. Our method begins by computing the ECC for tree structures, which are simpler representations of shapes. Subsequently, we introduce the concept of a 3D curve that corresponds to a smooth shape without holes, enabling the computation of the ECC for more complex shapes. By leveraging the 3D curve representation, our method provides an upper-bound approximation of the ECC, which can be effectively utilized in various applications. The proposed approach not only preserves the valuable properties of the ECC but also significantly reduces the computational burden, making it a more efficient and practical solution for graph-based representation and analysis of shapes in both 2D and 3D contexts.

Keywords: eccentricity transform · graph analysing · smooth shape · 3D curve · medial axis · distance transform

1 Introduction

The eccentricity transform (ECC) is a function that assigns to each point within a shape its geodesic distance to the furthest point [10]. In other words, it associates each point with the longest of the shortest paths connecting it to any other point within the shape [11]. The eccentricity transform is valuable for graph-based image analysis [1] due to its robustness to noise and minor segmentation errors [7], unique representation of a shape's geometry, and the ability to reveal connectivity and topology [9]. Additionally, it is useful for 2D and

Supported by the Vienna Science and Technology Fund (WWTF), project LS19-013.

M. Vento et al. (Eds.): GbRPR 2023, LNCS 14121, pp. 160–171, 2023.
https://doi.org/10.1007/978-3-031-42795-4_15

3D shape matching [5], enabling accurate comparisons between shapes [6]. Its invariance to translation, rotation, and scaling [5] allows for precise comparisons between shapes, while its capability to separate touching or overlapping objects enhances object detection and segmentation [12].

Calculation of the eccentricity transform for a shape can be computationally intensive. Using the *Dijkstra* algorithm, the complexity is $\mathcal{O}(n^2 \log(n))$ [5], where n is the number of vertices (pixels) in a 2D connected plane graph (2D image). However, under certain conditions, such as when a shape S has no holes, Ion [5] managed to reduce this complexity to $\mathcal{O}(|\partial S| \, n \log(n))$, where $|\partial S|$ is the number of vertices (pixels) on the boundary of the shape S. In this study, we examine basic shapes with increasing complexity, including line segments, tree structures , and smooth shapes. Our primary approach is to compute the eccentricity transform without the need for distance propagation. Nevertheless, when direct computation is not feasible, one can employ efficient parallel and hierarchical approaches [4] to expedite the propagation of distances.

Presently, our research focuses on the Water's Gateway to Heaven project[1], which involves high-resolution X-ray micro-tomography (μCT) and fluorescence microscopy. The image dimensions in this project exceed 2000 pixels per side, necessitating the use of the eccentricity transform to distinguish cells that are visually challenging to separate [2,3]. Consequently, fast computation of the eccentricity transform with low complexity is essential.

In this study, we begin in Sect. 3 by computing the eccentricity of line segments and extending the method to develop an efficient algorithm for tree structures. Next, in Sect. 4, we introduce the concept of a 3D curve for a shape and expand the proposed method to compute the eccentricity in smooth shapes without holes. Finally, Sect. 5 presents the simulations and results of our investigation.

2   Definitions

Basic definitions and properties of the ECC are introduced following [8,10]. Let the shape S be a closed set in $\mathbb{R}^2$ and ∂S be its border. A path π is the continuous mapping from the interval $[0, 1]$ to S. Let $\Pi(p_1, p_2)$ be the set of all paths between two points $p_1, p_2 \in S$ within the set S. The geodesic distance $d(p_1, p_2)$ between two points $p_1, p_2 \in S$ is defined as the length λ of the shortest path $\pi(p_1, p_2)$, such that $\pi \in S$, more formally

$$d(p_1, p_2) = min\{\lambda(\pi(p_1, p_2)) | \pi \in \Pi\} \qquad (1)$$

where

$$\lambda(\pi) = \int_0^1 \sqrt{1 + \dot{\pi}^2(t)} \, dt \qquad (2)$$

---

[1] https://waters-gateway.boku.ac.at/.

where $\dot{\pi}$ is a parametrization of the path from $p_1 = \dot{\pi}\ (0)$ to $p_2 = \dot{\pi}\ (1)$. The eccentricity transform of a simply connected planar shape S is defined as, $\forall p \in S$

$$ECC_S(p) = max\{d(p,q)|q \in S\} = max\{d(p,q)|q \in \partial S\} \tag{3}$$

i.e. to each point p it assigns the length of the shortest geodesics to the points farthest away from it. An *eccentric point* is defined as the point y that reaches a maximum in Eq. 3. Note that all eccentric points of a simply connected planar shape S lie on its border ∂S [10].

3   Tree Structure

A tree structure is an undirected graph characterized by its acyclic nature and connectedness, which means that there are no cycles and any two vertices are connected by exactly one path. A tree can be constructed by combining line segments that are connected together at branching points . Each line segment represents an edge in the tree, connecting two vertices, while the branching points serve as junctions where multiple line segments meet. By connecting line segments in this manner, it is possible to create a hierarchical structure with a single root node at the top and multiple branches extending downwards, ultimately forming a tree structure that captures the relationships and connectivity among the various nodes in the graph.

Consequently, computing the eccentricity transform in a tree can be achieved by considering the combination of its line segments. By analyzing each line segment's eccentricity and their connections at the branching points, the overall eccentricity transform for the entire tree structure can be determined. This approach simplifies the computation of the eccentricity transform for complex tree structures by breaking them down into smaller, more manageable segments, ultimately allowing for a more efficient calculation of the eccentricity values throughout the tree.

3.1   Line Segment

Consider a line segment, denoted by $l = (A, B)$, with endpoints A and B. In order to compute the eccentricity:

Proposition 1. *The eccentric points of a line segment are its corresponding two endpoints.*

Proof. Let us consider a line segment l with its two endpoints, A and B. Suppose that there exists a point $P \in l \backslash \{A, B\}$ such that Q is an eccentric point, meaning that Q is the farthest point away from P. If we move Q towards the corresponding endpoint, for instance point B, we obtain $\lambda(P, Q) < \lambda(P, B)$, which contradicts the original assumption. □

Proposition 2. *The eccentricity of a point P on a line segment $l = (A, B)$ is:*

$$ECC : P \in l \mapsto \mathbb{R}^+$$
$$ECC(P) = \max\{\lambda(A, P), \lambda(B, P)\} \tag{4}$$

where $\lambda(A, P)$ and $\lambda(B, P)$ represent the arc length of curves (A, P) and (B, P), respectively.

Proof. Based on Proposition.1, the two endpoints of a line segment are its eccentric points. Therefore, the eccentricity is the maximum value of these two endpoints to the point P (see Fig. 1a). □

It is important to note that the edges of a graph are not necessarily straight lines; in general, an edge can be a **curve**. Therefore, $\lambda(A, P)$ and $\lambda(B, P)$ typically represent the **geodesic distance** from point P to points A and B.

3.2   Branching Point

In a tree, vertices having only one incident edge are the leaves of the tree. We define a *branching point* as follows:

Definition 1 (Branching point). *A branching point in a tree is a vertex with a degree of more than two.*

Consider a tree consisting of a branching point B and k number of leaves.

Proposition 3. *The eccentricity of a point P in a tree T containing one branching point B and k leaves is:*

$$ECC(P) = \max\{\lambda(A, P), \lambda(B, P) + D_{max}\} \tag{5}$$

where $P \in l = (A, B)$ and D_{max} is the maximum distance of other leaves to the branching point B as follows:

$$D_{max} = \{\max\{\lambda(B, V_i)\}|\forall i \in k, V_i \neq A\} \tag{6}$$

Proof. With only one line segment $l = AB$, the eccentricity is computed based on Proposition 2. When adding another line segment BC that shares an endpoint with l, the eccentricity is computed as $ECC(P) = \max\{\lambda(A, P), (\lambda(P, B) + \lambda(B, C)\}$. To prove the proposition, we can iteratively connect a branch into the branching point B and keep the maximum distance as the result of the comparison to the previous branch (see Fig. 1b). By doing this, the $\lambda(B, C)$ is substituted with D_{max}. □

3.3   Tree

Let $T = (V, E)$ represent a tree comprising leaves and branching points. The attribute of an edge e is its *arc-length* $\lambda(e)$ where $e = (u, v) \in E$, $u, v \in V$, and $\lambda(e) = \lambda(u, v)$. The eccentricity calculation is performed using a hierarchical structure constructed over the input tree as we call it *hierarchical tree*. There are two types of movements in the hierarchical tree: *inward* and *outward*. The former is considered a **bottom-up** movement, while the latter is regarded as **top-down**.

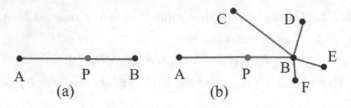

Fig. 1. computing the ECC. (a) a line segment, $ECC(P) = \lambda(A, P)$. (b) a branching point, $ECC(P) = \lambda(P, B) + \lambda(B, C)$.

Bottom-Up Movement. In this fashion, in order to compute the eccentricity of vertices, a stack of smaller reduced trees is constructed over the given input tree. Consider the input tree T, which serves as the base of the hierarchy. At each level k of this hierarchy, vertices are categorized into two types: leaf vertices and branching vertices. Let $\mathcal{L}_k$ represent the set of all leaf vertices at level k, and let $\mathcal{B}_k$ represent the set of all branching vertices at the same level. All vertices connected to a given vertex v (the adjacent vertices of v) are identified by $\mathcal{N}_k(v)$ at level k. In order to propagate distances to the upper levels of the hierarchy, an *intermediate distance*, $ID(v)$, is assigned to each vertex. Initially, all vertices have an intermediate distance of zero, i.e., $ID(v) = 0$ for all $v \in V$. A distance value for each branching point $b \in \mathcal{B}_k$ incident to at least one leaf is then calculated as follows:

$$D(b) = max\{\lambda(u, b) + ID(u) | \forall u \in \{\mathcal{N}_k(b) \cap \mathcal{L}_k\}\} \tag{7}$$

where $\mathcal{N}_k(b) \cap \mathcal{L}_k$ is a set of leaves that are adjacent to the branching point b. Subsequently, the leaves at the base level are contracted, leading to a smaller tree at the higher level, where the leaves of the smaller tree correspond to the branching points from the level below. This procedure is repeated, while the leaves are contracted in a bottom-up approach, ultimately leading us to the top of the hierarchy. At the top of the hierarchy, there is either one single vertex or two vertices. The longest path at the base of the hierarchical tree is the *diameter* of the tree, $dim(T)$.

Proposition 4. *The top of the hierarchical tree consists of a single vertex if and only if the length of the tree's diameter at the base level is an even number.*

Proof. Given that the tree's diameter is an even number, $dim(T)$ at the base level is expressed as $2k$. The process of leaf contraction at each level leads to a subsequent smaller tree T_1 at the upper level, where $dim(T_1)$ equals $(2k) - 2$. Upon repeated application of this reduction process at subsequent levels, we ultimately reach the apex, characterized by a single vertex where the diameter is equal to zero.

Proposition 5. *The top of the hierarchical tree consists of two vertices if and only if the length of the tree's diameter at the base level is an odd number.* □

Proof. Similar to the previous proof, here $dim(T) = 2k + 1$ at the base level. Through the contraction of leaves at each level, the resulting smaller tree at the upper level has $dim(T) = (2k + 1) - 2$. Therefore, repeating this reduction at upper levels leads to a tree with diameter 1 at the top which is a tree consisting of two vertices.

When a single vertex is present at the top, the computed distance value represents that vertex's eccentricity. However, when there are two vertices, labeled as v_1 and v_2, each with corresponding computed distance values D_1 and D_2, the eccentricity for these vertices is calculated as follows:

$$D_1 > D_2 \Rightarrow ECC(v_1) = max(D_1, D_2 + \lambda(v_1, v_2)) \ and \ ECC(v_2) = D_1 + \lambda(v_1, v_2)$$
$$D_1 = D_2 \Rightarrow ECC(v_1) = ECC(v_2) = D_1 + \lambda(v_1, v_2)$$
$$(8)$$

Top-Down Movement. The eccentricities of remaining vertices are iteratively computed in a top-down fashion. The tree at the top is successively expanded through outward movement until it reaches the base of the hierarchy, where each vertex is assigned its corresponding eccentricity value.

Consider a hierarchical tree with a single vertex at the top level $k+1$. Through a bottom-up approach, the eccentricity of the top vertex is determined based on Eq. (7) by taking the maximum value from the sum of intermediate values and the arc lengths of each leaf at level k. Let v_m be the leaf at level k that corresponds to the maximum value. Additionally, let $D'(b)$ be the maximum value of b where v_m is eliminated from its adjacency:

$$D'(b) = max\{\lambda(u, b) + ID(u) | \forall u \in \{\mathcal{N}_k(b) \cap \mathcal{L}_k \backslash v_m\}\} \qquad (9)$$

Employing a top-down approach, the eccentricities of the remaining vertices are iteratively computed. A leaf vertex at level $k + 1$, whose eccentricity has already been determined, transmits its eccentricity to the corresponding branching point b at the lower level k. The eccentricity of each leaf $v \in \mathcal{L}_k, \backslash v_m$ at level k is computed as follows:

$$ECC(v) = \lambda(v, b) + ECC(b), \quad v \in \mathcal{L}_k, \ b \in \mathcal{B}_k \qquad (10)$$

To calculate the eccentricity of the leaf vertex v_m, a comparison is made between the value derived from Eq. (9) and the value of $ECC(b) - \lambda(v, b)$. Subsequently, the eccentricity of the vertex v_m is computed as follows:

$$ECC(v_m) = max\{D'(v_m) + \lambda(v_m, b), \ ECC(b) - \lambda(v_m, b)\} \forall b \in \{\mathcal{N}_k(b) \cap \mathcal{L}_k\}$$
$$(11)$$

Figure 2a shows an instance of a hierarchical tree featuring three levels and a single vertex at its apex. The bottom-up movement is depicted on the left side, whereas the top-down progression is illustrated on the right. Intermediate distances are visually represented within a box, while the eccentricity of vertices is

---

Algorithm 1. Computing the eccentricity (ECC) in a Tree

1: **Input:** Tree: $T = (V, E)$, $\mathcal{L}_k$: set of leaves at level k, $\mathcal{B}_k$: set of branching vertices
   at level k, $ID(v)$: Intermediate Distance of v , k: level of the hierarchy , $D(b)$:
   distance value for a branching point , $\lambda(u, v)$: arc length of edge $e = (u, v)$
2: Initialization: $ID(v) = 0 \,\forall v \in V$, $\quad$ k $= 1$
3: **While** $\exists \, v \in \mathcal{L}_k$ (**bottom-up** movement in the tree hierarchy)
4: $\quad D(b) = max\{\lambda(u, b) + ID(u) | \forall u \in \{\mathcal{N}_k(b) \cap \mathcal{L}_k\}\}$
5: $\quad k = k + 1$
6: **end**
7: $ECC(b) = D(b) \quad$ (Top of the hierarchy)
8: $k = k - 1$
9: **While** $k > 0$ (**top-down** movement in the tree hierarchy)
10: $ECC(v) = \lambda(u, v) + ECC(u)$, $\quad v \in \mathcal{L}_k$, $\; u \in \mathcal{B}_k$
11: **end**

---

denoted by a number enclosed in a red circle. In the event that two vertices reside
at the top following the computation of their eccentricities, the calculation of the
eccentricity for the remaining vertices aligns with the methodology previously
described (see Fig. 2b). The specifics of this method are outlined in Algorithm 1.
The algorithm's complexity is determined by the number of branching points in
the tree.

4   Shape

Calculating the eccentricity of trees can potentially enable us to extend the
proposed method for more complex shapes. In a tree, the leaves are recognized
as the eccentric points of the structure. However, what constitutes the eccentric
points in an arbitrary shape? If we are unable to identify the eccentric points,
is it possible to at least estimate them and compute the eccentricity of a shape?
To address this question, we propose the following method, which may offer an
upper bound for the eccentricity value of a given shape.

4.1   3D Curve of a Shape

The medial axis (MA) of a shape is a collection of center points of all maximally
inscribed circles (or spheres in 3D) . These circles touch the shape's boundary
at two or more points, with their centers forming the MA, also known as the
topological skeleton. This axis captures connectivity of the shape, providing
a compact and informative representation. Conversely, the distance transform
assigns a value to each point within the shape, representing the shortest distance
from that point to the shape's boundary.

The proposed method combines the MA and distance transform to effectively
reconstruct the original shape. First, the radius of the maximally inscribed circle
(or sphere in 3D) is obtained for each point on the MA using distance transform
values. Then, these circles (or spheres) grow at each point on the MA, and

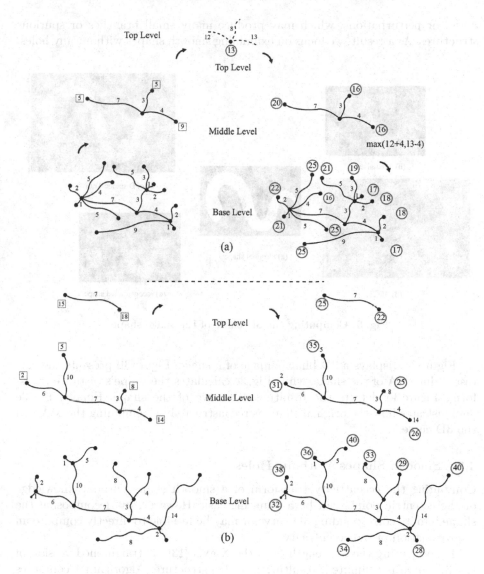

Fig. 2. Computing the *ECC* in a hierarchical tree. (a) One vertex and (b) two vertices at the top level.

their union is taken to reconstruct the shape. The reconstructed shape may not be an exact replica of the original, particularly if derived from a noisy or imperfect representation. However, it preserves the shape's essential topology and connectivity, providing a reasonable approximation.

In this paper, a shape is represented by its MA and corresponding distance transform values, resulting in what we refer to as the **3D curve** of the shape. The MA can be sensitive to non-smooth shapes or shapes with small irregularities, noise, or perturbations, which may produce many small branches or spurious structures. As a result, we focus on examining smooth shapes without any holes.

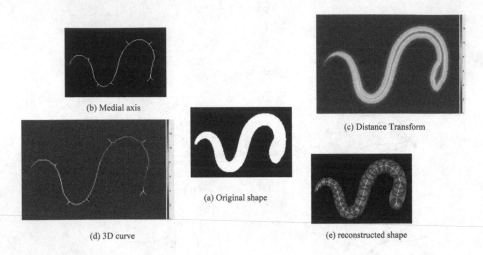

Fig. 3. Computing the 3D curve of the snake shape.

Figure 3a displays a 2D binary image of a snake. Figure 3b presents the corresponding MA of the snake, while Fig. 3c calculates the shape's distance transform. Figure 3d depicts the resulting 3D curve of the shape. Finally, Fig. 3e demonstrates how the original shape is reconstructed by combining the MA and the 3D curve.

4.2 Smooth Shapes Without Holes

Computing the eccentricity transform of a smooth shape without knowledge of the eccentric points can be a daunting task. However, by decomposing the shape into its corresponding 3D curve, it may be feasible to directly compute an approximation of the eccentricity.

By projecting the arc length onto the X-axis [13], a straightened version of the 3D curve is computed, resulting in a tree structure. Algorithm 1 computes the ECC of the MA. For the remaining points not on the MA of the shape, each point of the shape computes its corresponding distance to the MA. Afterward, the eccentricity of the resulting point is computed along the MA (geodesic distance) to find the corresponding eccentric point on the MA. Finally, the distance transform of the computed eccentric points is added to the previous distances. However, due to the concavity of a shape, the computed eccentricity using the proposed method is generally an overestimate of the true eccentricity of the original shape. This is because the method computes the geodesic distance along the MA, and the concavity of the shape can lead to the distance being overestimated in some regions [10]. As a result, the computed eccentricity represents an upper-bound for the eccentricity transform of the smooth shape.

5   Simulation and Result

The effectiveness of the proposed method for computing the eccentricity transform was evaluated through a simulation of a snake shape, as depicted in Fig. 4. The medial axis of the shape was first computed, and for each point on the medial axis, its corresponding eccentric point was computed (see Fig. 4a). The color of each point in Fig. 4a corresponds to the value of its corresponding eccentric point. The thickness of the smooth shape was then determined using the distance transform (Fig. 4b). The resulting upper bound of the eccentricity transform is presented in Fig. 4c, while the ground truth was computed and shown in Fig. 4d. Table 1 shows the computational error by comparing the ground truth with the upper bound of the eccentricity.

The presented results demonstrate that the proposed method offers a promising approach for achieving more accurate eccentricity computation. Notably, the method is capable of being computed with $\mathcal{O}(b)$ complexity when b is the number of branching points of the tree of the medial axis.

Table 1. Comparing the ECC of ground truth with the computed upper bound.

Mean Absolute Error(MAE)	Relative MAE	Mean Square Error (MSE)	Relative MSE
1.1832	0.0348	10.9440	0.0025

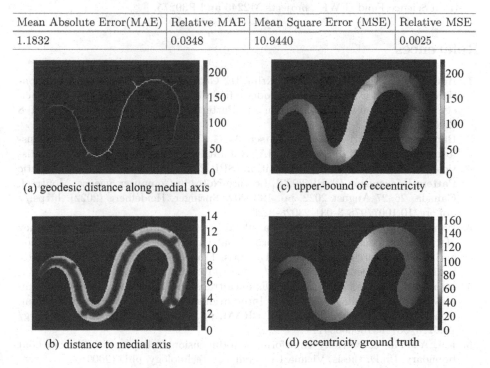

(a) geodesic distance along medial axis

(c) upper-bound of eccentricity

(b) distance to medial axis

(d) eccentricity ground truth

Fig. 4. Computation of the eccentricity transform.

6   Conclusion

This paper introduces an innovative approach for computing the eccentricity transform of a tree. The proposed method achieves $\mathcal{O}(n)$ complexity, where n is the number of branching points. By utilizing the introduced 3D curve representation, the paper extends the method to compute the eccentricity of smooth shapes without holes. This allows for a faster computation of an upper bound for the eccentricity, which is useful in many applications in 2D and 3D shape analysis, such as shape matching, classification, and recognition. The main result of this paper demonstrates that the proposed algorithm provides a reliable estimation of the actual eccentricity, and it closely approximates the ground truth. Moreover, the reduced computational complexity of the proposed approach promises efficient processing of more complex shapes in future work, which is crucial for real-world applications where computational resources and time are limited.

Acknowledgment. We acknowledge the Paul Scherrer Institut, Villigen, Switzerland for the provision of beamtime at the TOMCAT beamline of the Swiss Light Source and would like to thank Dr. Goran Lovric for his assistance. This work was supported by the Vienna Science and Technology Fund (WWTF), project LS19-013, and by the Austrian Science Fund (FWF), projects M2245 and P30275.

References

1. Aouada, D., Dreisigmeyer, D.W., Krim, H.: Geometric modeling of rigid and non-rigid 3D shapes using the global geodesic function. In: 2008 IEEE Computer Society Conference on Computer Vision and Pattern Recognition Workshops, pp. 1–8 (2008)
2. Banaeyan, M., Carratù, C., Kropatsch, W.G., Hladůvka, J.: Fast distance transforms in graphs and in gmaps. In: IAPR Joint International Workshops on Statistical Techniques in Pattern Recognition (SPR 2022) and Structural and Syntactic Pattern Recognition (SSPR 2022), Lecture Notes in Computer Science, Montreal, Canada, 26–27 August 2022, pp. 193–202. Springer, Heidelberg (2022). https://doi.org/10.1007/978-3-031-23028-8_20
3. Banaeyan, M., Kropatsch, W.G.: Parallel $\mathcal{O}(log(n))$ computation of the adjacency of connected components. In: International Conference on Pattern Recognition and Artificial Intelligence (ICPRAI), Paris, France, 1–3 June 2022, pp. 102–113. Springer, Heidelberg (2022). https://doi.org/10.1007/978-3-031-09282-4_9
4. Banaeyan, M., Kropatsch, W.G.: Distance transform in parallel logarithmic complexity. In: Proceedings of the 12th International Conference on Pattern Recognition Applications and Methods - ICPRAM, pp. 115–123 (2023). https://doi.org/10.5220/0011681500003411
5. Ion, A.: The eccentricity transform of n-dimensional shapes with and without boundary. Ph.D. thesis, Vienna University of Technology, phD (2009)
6. Ion, A., Artner, N.M., Peyré, G., Kropatsch, W.G., Cohen, L.: Matching 2D & 3D articulated shapes using the eccentricity transform. Comput. Vision Image Underst. **115**(6), 817–834 (2011)

7. Ion, A., Peltier, S., Alayranges, S., Kropatsch, W.G.: Eccentricity based topological feature extraction. In: Alayranges, S., Damiand, G., Fuchs, L., Lienhardt, P. (eds.) Workshop Computational Topology in Image Context, CTIC 2008. Universit'e de Poitiers (2008)
8. Ion, A., Peyré, G., Haxhimusa, Y., Peltier, S., Kropatsch, W.G., Cohen, L.: Shape matching using the geodesic eccentricity transform - a study. In: Ponweiser, W., Vincze, M. (eds.) Proceedings of 31st OEAGM Workshop, pp. 97–104. Österreichische Computer Gesellschaft (2006). Band 224
9. Janusch, I., Kropatsch, W.G.: Lbp scale space origins for shape classification. In: Artner, N.M., Janusch, I., Kropatsch, W.G. (eds.) Proceedings of the 22nd Computer Vision Winter Workshop 2017, pp. 1–9. TU Wien, PRIP Club (2017). iSBN 978-3-200-04969-7
10. Kropatsch, W.G., Ion, A., Haxhimusa, Y., Flanitzer, T.: The eccentricity transform (of a digital shape). In: Kuba, A., Nyúl, L.G., Palágyi, K. (eds.) DGCI 2006. LNCS, vol. 4245, pp. 437–448. Springer, Heidelberg (2006). https://doi.org/10.1007/11907350_37
11. Kropatsch, W.G., Ion, A., Peltier, S.: Computing the eccentricity transform of a polygonal shape. In: Rueda, L., Mery, D., Kittler, J. (eds.) CIARP 2007. LNCS, vol. 4756, pp. 291–300. Springer, Heidelberg (2007). https://doi.org/10.1007/978-3-540-76725-1_31
12. Ma, J., et al.: How distance transform maps boost segmentation cnns: an empirical study. In: Arbel, T., Ben Ayed, I., de Bruijne, M., Descoteaux, M., Lombaert, H., Pal, C. (eds.) Proceedings of the Third Conference on Medical Imaging with Deep Learning. Proceedings of Machine Learning Research, vol. 121, pp. 479–492. PMLR (2020)
13. Pucher, D., Artner, N.M., Kropatsch, W.G.: 2D tracking of platynereis dumerilii worms during spawning. In: Čehovin, L., Mandeljc, R., Štruc, V. (eds.) Proceedings of the 21st Computer Vision Winter Workshop 2016, pp. 1–9 (2016)

Graph-Based Deep Learning on the Swiss River Network

Benjamin Fankhauser[1]($\boxtimes$)(iD), Vidushi Bigler[2](iD), and Kaspar Riesen[1](iD)

[1] Institute of Computer Science, University of Bern, Bern, Switzerland
{benjamin.fankhauser,kaspar.riesen}@unibe.ch
[2] Institute for Optimisation and Data Analysis, Bern University of Applied Sciences,
Biel, Switzerland
vidushi.bigler@bfh.ch

Abstract. Major European rivers have their sources in the Swiss Alps. Data from these rivers and their tributaries have been collected for decades with consistent quality. We use GIS data to extract the structure of each river and link this structure to 81 river water stations (that measure both water temperature and discharge). Since the water temperature of a river is strongly dependent on the air temperature, we also include 44 weather stations (which measure, for instance, air or soil temperature). Based on this large data corpus, we present in this paper a novel graph representing the water network of Switzerland. Our goal is to accelerate the research of the complex relationships at the (Swiss) water bodies. In particular, we present different graph-based pattern recognition tasks that can be solved on the novel water body graph. In a first evaluation, we use graph-based methods to solve two of these tasks, outperforming current state-of-the-art systems by several percentage points.

Keywords: Water body graph · Water temperature · LSTM · Recurrent Neural Network · Graph data

1 Introduction

Water temperature – an important variable in our ecosystem – is mainly influenced by air temperature. That is, on the water surface a direct exchange with the surrounding air takes place. Thereby, solar radiation is either absorbed by particles in the water or the river bed, then transformed to heat and finally exchanged with the water. Other factors that influence the temperature of water bodies are snow melting, rain, ground water inflow, but also the rate of discharge. Last but not least, also human-made infrastructure plays a pivotal role. For instance, the climate regime shift (CRS) in the late 1980s, caused by anthropogenic and natural origin, led to a sudden increase in water temperature [1,2].

Supported by Swiss National Science Foundation (SNSF) Grant Nr. PT00P2_206252. Data are kindly provided by the Federal Office for the Environment and MeteoSwiss.

Three major European rivers have their sources in Switzerland, namely Rhine, Rhône and Inn. In addition, the river Ticino rises in Switzerland, which contributes significantly to the river Po. Furthermore, a large part of the Alps, which separate the southern and northern climatic zones, lies in Switzerland. The Alps are in turn home to large glaciers and huge snow reservoirs, as well as human-made infrastructure such as power plants and dams. In addition, the topology of Switzerland also consists of hilly lowlands, where small rivers flow slowly and are influenced by both large cities and agriculture. Further towards the borders of Switzerland we have the big rivers which are less affected by small disturbances. There are also some medium sized lakes where the inflowing water stays for a long time and thus the outflowing water is only slightly influenced by the inflowing water (there the surface temperature is mainly influenced by the exchange in the atmosphere). Overall, we find that in Switzerland there is a fascinating network of water bodies that has a high complexity.

The present paper is concerned with water temperature predictions using air temperatures by means of graph-based pattern recognition and machine learning. Actually, concerning the climate crisis, rising water temperatures will have a big impact on the Swiss ecosystem. For instance, certain species of fish will not be able to reproduce anymore when the water temperature reaches a certain threshold [3]. Different climate models exist that project air temperature for the future in various versions [4]. Thus, our hypothesis is that it is rewarding to explore more accurate modelling of the air-water model, as this will also lead to better long-term projections of the water temperature.

The contribution of the present paper is threefold. First, based on data of the water bodies stemming from a Geographic Information System (GIS), as well as decades of measurements of dozens of stations, we create a novel and large-scale graph that aims to comprehensively capture and model the complexity of the Swiss water network. Hence, the basis of our research is similar in spirit to other important prediction tasks such as analyses of transportation networks [5], or predictions of loads on networks of power grids [6], to name just two examples. Second, the novel graph allows us to reconsider current approaches to predicting water temperature in rivers. We propose different tasks related to water temperature prediction that can potentially be solved with graph-based pattern recognition algorithms. Third, for two of these tasks, we propose a graph-based prediction system and show that this novel system significantly outperforms two current state-of-the-art methods.

The remainder of this paper is organized as follows. In Sect. 2, we describe two state-of-the-art methods that are currently used for water temperature prediction, viz. the Air2Stream method [7] as well the adaptation of LSTM neural networks [8]. In Sect. 3, we thoroughly describe the novel graph that models the Swiss water bodies and introduce the challenges in predicting water temperature. The novel method for water temperature prediction that employs a graph-based model is then presented and evaluated later in Sect. 4. Finally, in Sect. 5, we draw conclusions and propose possible rewarding avenues for future research activities.

2   Related Work

We are not the first to attempt to predict water temperatures based on air temperatures. In the following two subsections, we present two state-of-the-art models that are actually used as reference systems in our empirical evaluation.

2.1   Air2Stream

Air2Stream is a physically inspired model of the relationship between air and water temperature based on air temperature and discharge [7]. Basis of this model is a differential equation linearized using a Taylor series expansion. The resulting equation has eight tuneable parameters which are calibrated using training data. In particular, the original method employs a particle based optimisation scheme for training and is quite sensitive to the chosen hyper-parameter. In the present paper we use the predictions presented in [9].

2.2   LSTM on Water Data

Long short-term memory (LSTM) is a special type of a recurrent neural network (RNN) [10]. An RNN is a neural network that is applied to a time series on every time step. In addition, an LSTM keeps track of a hidden state and a memory state, two vectors which are inputs to the next time step and will be altered by the LSTM. Thus, the resulting backpropagation variant is called backpropagation through time [11]. To have a trade off between the time series and the update steps, one works with a certain time window, where at the end of every window the gradients are computed and an update step is made. LSTMs have been particularly designed to encounter the vanishing gradient problem. This problem occurs when the back propagation through time has to overcome a lot of time steps and repeated multiplications tend to unstable numeric conditions. In general LSTMs show state-of-the-art results on various time series data [12], and can be applied to the task of water temperature prediction [8,13] or water level (discharge) prediction [14,15].

3   The Swiss Water Body Graph

3.1   Construction of the Graph

One of the major contributions of the present paper is that we provide a novel graph based on the Swiss water body. We construct a knowledge graph containing information about the location of river beds (from a GIS), weather data of 44 weather stations (air temperature and more atmospheric measurements), and water data of 81 water stations (water temperature and discharge).

The knowledge Graph $G = (V, E, a_V, a_E)$ is a graph with nodes V and edges E. Each node $v \in V$ has an assigned type T_v. Currently we have three types of nodes i.e. $T_v \in \{\text{water station}, \text{universal river node}, \text{weather station}\}$. The universal river node is used to model the river itself with sources or river mouths.

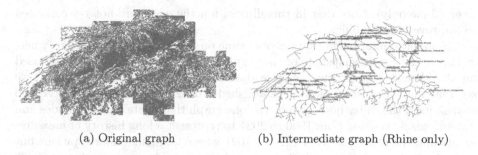

(a) Original graph   (b) Intermediate graph (Rhine only)

Fig. 1. (a) Original graph before pruning (all edges represent water). (b) Subgraph representing the river Rhine after pruning and added water stations. The illustrated graph is a tree where the water of a child station flows to its parent station.

There is currently only one edge type, which models the connectivity of the nodes. The functions $a_V : V \rightarrow \mathbb{R}^n$ and $a_E : E \rightarrow \mathbb{R}^m$ deliver additional attributes to nodes and edges. Function a_E assigns the edge length in meters to each edge. The nodes are attributed by the air temperature, or water temperature and discharge (depending on the actual type T_v of the node). The data basis for these attributes is thoroughly preprocessed. In particular the data is min-max normalised and outliers are removed by the Federal Office for the Environment (FOEN) as part of their quality control. Each weather station is manually connected by means of an edge to $n \geq 1$ water stations as proposed in [9].

At first glance, the considered data basis seems of natural origin, yet it is not. The current rivers are the product of decades of human intervention of stratification, city planning, power plants, and renaturation. Also the placement and running of the water and weather stations are obviously human based decisions and can change in future. Keeping a constant and high quality of measurements is challenging as it requires decades of stability in the corresponding country, which fortunately is the case in Switzerland.

The original GIS graph contains 258,103 edges and 258,191 nodes representing different types of river segments as well as lake contours. We apply the following preprocessing on this original graph. First, we prune the leaf nodes as not every side creek is important. Nodes that actually contain water stations are never purged. Then, we run a spanning tree algorithm in order to find the shortest paths of the water flow and remove any ambiguity in the graph (for example, when both sides of a lake are modelled as two edges in the graph). Finally, we collapse all edges such that only the connectivity between water stations is left in the form of a tree. This process allows us to compress edge information like river bed length as sum of all collapsed segments between two water stations.

This resulting graph consists of four trees (representing the rivers Rhine, Rhône, Inn, Ticino) with a total of 73 nodes and 69 edges. In Fig. 1(a) the original graph and in Fig. 1(b) the resulting graph is shown (representing the river Rhine only). Note that in this illustration the edges are not yet collapsed to improve visualisability.

As we have man made changes over time on the underlying water body network, we can create snapshots of the graph at every useful point in time. Based on the visualisation of the available data from 1980 to 2021 (see Fig. 2), we see that many new stations were established after 2002. Hence, we propose two snapshots of the water body graph, viz. one graph that contains fewer nodes and measurements ranging from 1990 to 2021 to represent a long history of measurements and the other one from 2010 to 2021 where we include more stations but on a shorter period of time. In both snapshots we apply an approximate 80/20 training-test split at the end of 2014 and 2017, respectively. The two graphs are named G_{1990} and G_{2010} from now on. Both graphs will be made publicly available for research purpose on the Git Repository of our research group[1].

Fig. 2. Visualisation of the available data from 1980 to 2021. Each row represents one station. Dark grey pixels indicate that at a certain day the river water temperature, discharge and air temperature are available. Light grey pixels indicate that at least one of the three values is missing.

3.2   Proposed Water Challenges

The novel graph based representation defined above allows us to rethink current approaches for water temperature prediction. We propose five different benchmark tasks, that can potentially be solved on the basis of the novel graph.

Task 1 - Model Air Temperature Relationship: In this task the goal is to model the relationship between the air and water temperature. This challenge has already been extensively studied and it is what models like Air2Stream [7] or LSTMs [8] are aiming at. Formally, we have both air temperature $a_0, ..., a_t$, the discharge $q_0, ..., q_t$ of $t + 1$ time steps and the goal is to find a model f that predicts the water temperature w_t at time t: $f(a_0, ..., a_t, q_0, ..., q_t) = w_t$.

Task 2 - k-Day Forecast: In this task, we do not have access to same day measurements anymore. Given the air temperature $a_0, ..., a_t$ and the discharge $q_0, ..., q_t$ of $t+1$ time steps, the goal is to predict the water temperature w_{t+k} in k days (we define $k \in \{3, 7, 30\}$). Formally, we seek a model $f(a_0, ..., a_t, q_0, ..., q_t) = w_{t+k}$. Obviously, the larger k is choosen, the harder the problem (setting $k = 0$ results in Task 1).

Task 3 - Recover from Neighbours: Each water station is built at a certain construction time b_t. One problem of our graph is missing data for this station at times $t < b_t$. The goal of this task is to learn the data of a node for time points

---

[1] https://github.com/Pattern-Recognition-Group-UniBe/swiss-river-network

$t < b_t$ based on the relationships with its neighbours. By filling in missing data, this procedure allows us to construct an estimated graph of water temperatures back to 1980 using all stations (although we cannot assess the quality of the estimates).

Task 4 - Work on Degenerated Data: A challenge for any sensing and recognition system are degenerated sensors. The fourth task is to detect and repair potentially corrupted data. Formally, we define a function drift $d(n) \in \mathbb{R}$, where n is the n-th day after construction and d is a function to model the amount of drift. The drift is then added to the water temperature during training: $w'_t = w_t + d(t - b_t)$, where w'_t is the degenerate training data and b_t is the construction time of the water station.

Task 5 - Few Shot Learning: The goal of this task is to minimise the effort required to collect water temperatures. Imagine a mobile sensor system that is moved from one place to another every month. When the mobile sensor system is on site, the data is available and can be used for training. The goal is to have as few of these mobile sensor systems in use as possible and still achieve a reasonable estimation of the water temperatures.

4   Proposed Method and Experimental Evaluation

4.1   Experimental Setup and Reference Models

In this paper, we use the snapshots of the graphs G_{1990} and G_{2010} as described in Sect. 3.1 and the corresponding training and test splits to solve Task 1 and Task 2 as defined in Sect. 3.2 (that is, predicting the water temperature in k days with $k \in \{0, 3, 7, 30\}$). To investigate the quality of the prediction, we measure and report widely used metrics, namely the Root Mean Squared Error (RMSE) and the Mean Absolute Error (MAE) on the test set. In addition, we measure and report the Nash-Sutcliffe model Efficiency Coefficient (NSE), which is often used to assess the predictive skill of hydrological models. Formally, the three ratios are defined as follows

$$\text{RMSE} = \sqrt{\frac{1}{n} \sum_{i=1}^{n} (y_i - \hat{y}_i)^2} \tag{1}$$

$$\text{MAE} = \frac{1}{n} \sum_{i=1}^{n} |y_i - \hat{y}_i| \tag{2}$$

$$\text{NSE} = 1 - \frac{\sum_{i=1}^{n} (y_i - \hat{y}_i)^2}{\sum_{i=1}^{n} (y_i - \bar{y})^2} \tag{3}$$

where n describes the number of measurements, y_i the actual measured value, $\hat{y}_i$ the value estimated by the model, and $\bar{y}$ the mean of the actual measured values. For a perfect model with an estimation error variance equal to zero, the resulting NSE equals 1. That is, values of the NSE nearer to 1 suggest a model with more

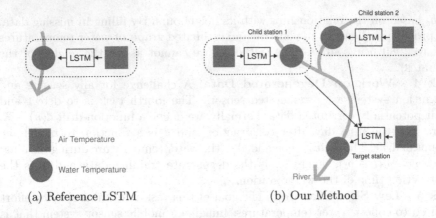

(a) Reference LSTM   (b) Our Method

Fig. 3. (a) The reference method models the air to water relationship in a 1-to-1 manner [8]. (b) The proposed method makes use of the local neighbourhood on the novel water body graph. We adapt our LSTM architecture to the amount of child stations contributing to the target station and train one such LSTM per target station.

predictive skill. While for the errors, of course, values closer to 0 indicate good prediction quality.

For our evaluation, we use a total of three different reference models.

1. Air2Stream: The Air2Stream model as presented in Sect. 2.1. We only provide here the RMSE results form [9] (and we cannot use the other metrics for comparison).
2. Baseline: The baseline system refers to the unweighted average of the water temperature of the target station and the water temperatures of its child stations using the reference LSTMs (see below).
3. LSTM: For this reference system [8], we use LSTMs that take the air temperature as input (as described in Sect. 2.2 – see Fig. 3(a)). To find a suitable architecture, we perform a grid search on the width of the hidden layers, the depth of the LSTM, the learning rate, and the weight decay (we use the Adam optimiser). During validation we obtain the best results for 32 in width, 1 in depth, 0.01 for learning rate, and a weight decay of 1e-6.

4.2   The Novel Graph-Based Model

For our new model, we use the four graphs described in detail in Sect. 3.1. The new model uses the locality of the graph structure to model the time series data and consists of two different nodes.

– Child station: Water station upstream to the target station
– Target station: Water station we want to predict.

We extract a subgraph for each target station with its c child stations. One such subgraph with one target station and $c = 2$ child stations is shown in Fig. 3(b).

In Task 1 and Task 2, we do not have access to measured water temperatures of the child stations as input, but we can estimate them using any air to water model. For each child station, we train a reference LSTM to obtain an estimate of the water temperature. Then we train an additional LSTM for the target station. This LSTM is given the estimated water temperatures of the child stations and the air temperature of the target station as input (see Fig. 3(b)).

More formally, the resulting recurrent neural network consists of an LSTM layer with the input size $c+1$. The LSTM uses a larger hidden space than its input size. The size of the hidden space is determined by a factor of the input size. After the LSTM layer, we project the hidden space to the desired output size using a linear layer. Our neural network models the function $f(\hat{w}_t^{(1)}, ..., \hat{w}_t^{(c)}, a_t^{(ts)}) = \hat{w}_{t+k}^{(ts)}$ where $\hat{w}_t^{(x)}$ is the estimated water temperature at child station x and $a_t^{(ts)}$ is the air temperature at the target station ts at time t, and k depends on the current prediction task ($k \in \{0, 3, 7, 30\}$).

For the training of our model, we perform a grid search for both width and depth of the LSTM and use the Adam optimiser with a learning rate of 0.01 and a weight decay of 1e-6.

Graph Neural Networks (GNNs) with message passing [16] are somehow related to the proposed method. Similar in spirit is, for instance, Graph-SAGE [17]. However, while GraphSAGE uses an LSTM to handle a flexible amount of neighbours during the message aggregation phase, we have a fixed amount of child stations but a flexible amount of time steps to handle. Moreover, GNNs aim to process the graph as a whole input unit. In the proposed method we train a neural network individually per target station. This removes any inductive property as our trained networks do not generalise to other graphs.

4.3   Test Results

The results we obtain on both versions of the graph (i.e. G_{1990} and G_{2010}) are shown in Table 1. The metrics RMSE, MAE, and NSE are reported for the respective test years. In column $k = 0$, the results for estimations of the same day are shown (Task 1). In the columns $k = 3$, $k = 7$, and $k = 30$, we show the prediction results for 3, 7, and 30 days in the future, respectively (Task 2).

First, we observe that our new model performs best for both graphs and all four tasks (measured across all three evaluation metrics). On average, we outperform the state-of-the-art method Air2Stream by 23%. Moreover, on average, the novel location-based method outperforms the state-of-the-art LSTMs by about 5%, remarkably more at the most difficult task $k = 30$. The results of the baseline show that a simple average of locally connected water temperatures is a poor estimate.

Regarding the results, we conclude that the proposed method is a flexible extension to any system that models the relationship of air temperature and

Table 1. The results achieved on the test sets by our method and the reference systems on two versions of the graph (G_{1990} and G_{2010}). In the $k = 0$ column, we report the results for the same day relation (Task 1), and in the $k = 3$, $k = 7$, and $k = 30$ columns the forecasts for 3, 7, and 30 days in future, respectively (Task 2). The best result per metric, task and graph is shown in bold face. *The Air2Stream model uses similar years for training but a different set of test years.

Metric	Method	Graph Version							
		G_{1990}				G_{2010}			
		$k=0$	$k=3$	$k=7$	$k=30$	$k=0$	$k=3$	$k=7$	$k=30$
RMSE	Air2Stream*	1.05	-	-	-	1.05	-	-	-
	Baseline	1.90	2.00	2.16	3.02	1.65	1.87	2.00	2.41
	Reference LSTM	0.80	1.10	1.37	2.29	0.91	1.24	1.43	2.24
	Ours	**0.75**	**1.07**	**1.30**	**1.77**	**0.85**	**1.19**	**1.39**	**1.65**
MAE	Baseline	1.58	1.65	1.79	2.38	1.36	1.53	1.62	1.95
	Reference LSTM	0.60	0.84	1.04	1.68	0.69	0.93	1.09	1.69
	Ours	**0.56**	**0.81**	**0.99**	**1.36**	**0.63**	**0.89**	**1.05**	**1.26**
NSE	Baseline	0.82	0.81	0.77	0.54	0.85	0.83	0.81	0.76
	Reference LSTM	**0.97**	0.94	0.90	0.74	0.96	0.93	0.91	0.77
	Ours	**0.97**	**0.95**	**0.93**	**0.86**	**0.97**	**0.94**	**0.92**	**0.89**

water temperature. We argue that our system is able to capture water temperature changes of upstream stations, which in general results in an improvement of the prediction accuracy. A more in depth analysis of the performance of individual stations, however, also reveals that there is no improvement for some individual stations. The reason for this observation is that some water stations have no dependence on their upstream water stations (e.g., when a lake lies between two stations).

5   Conclusion and Future Work

In this paper, we address the difficult task of analysing water networks in complex environments. This is indeed an important task, as the climate crisis is one of the greatest challenges facing humanity. We propose to model the complex water network of Switzerland using a graph. Based on this graph, we propose five different challenging tasks that can potentially be solved using graph-based pattern recognition or machine learning methods. Two of these five tasks are solved in this paper using a graph-based model built on LSTMs. In a large-scale experimental evaluation, we show that the proposed model can improve the widely used Air2Stream model by about 23% and an isolated (i.e., non-graph-based) LSTM by about 5% We see many worthwhile future research activities. Currently we are working with the authorities to extend the graph with more water stations as well as other node types like cities, power plants and lakes. Moreover, we will tackle the remaining benchmark tasks and explore more possibilities of neural networks on our novel graph.

References

1. Reid, P.C., et al.: Global impacts of the 1980s regime shift. Glob. Change Biol. **22**(2), 682–703 (2016)
2. Woolway, R.I., Dokulil, M.T., Marszelewski, W., Schmid, M., Bouffard, D., Merchant, C.J.: Warming of central European lakes and their response to the 1980s climate regime shift. Clim. Change **142**, 505–520 (2017)
3. Dahlke, F.T., Wohlrab, S., Butzin, M., Pörtner, H.O.: Thermal bottlenecks in the life cycle define climate vulnerability of fish. Science **369**(6499), 65–70 (2020)
4. CH2018: Climate scenarios for Switzerland. Technical report, National Centre for Climate Services, Zurich (2018)
5. Yu, B., Yin, H., Zhu, Z.: Spatio-temporal graph convolutional networks: a deep learning framework for traffic forecasting. arXiv preprint arXiv:1709.04875 (2017)
6. Khodayar, M., Liu, G., Wang, J., Khodayar, M.E.: Deep learning in power systems research: a review. CSEE J. Power Energy Syst. **7**(2), 209–220 (2020)
7. Toffolon, M., Piccolroaz, S.: A hybrid model for river water temperature as a function of air temperature and discharge. Environ. Res. Lett. **10**(11), 114011 (2015)
8. Qiu, R., Wang, Y., Rhoads, B., Wang, D., Qiu, W., Tao, Y., Wu, J.: River water temperature forecasting using a deep learning method. J. Hydrol. **595**, 126016 (2021)
9. Råman Vinnå, L., Zappa, M., Piccolroaz, S., Bigler, V.C., Epting, J.: Swiss-wide future river temperature under climate change, swissfurite (unpublished)
10. Hochreiter, S., Schmidhuber, J.: Long short-term memory. Neural Comput. **9**(8), 1735–1780 (1997)
11. Werbos, P.J.: Backpropagation through time: what it does and how to do it. Proc. IEEE **78**(10), 1550–1560 (1990)
12. Siami-Namini, S., Tavakoli, N., Namin, A.S.: A comparison of ARIMA and LSTM in forecasting time series. In: 2018 17th IEEE International Conference on Machine Learning and Applications (ICMLA), pp. 1394–1401. IEEE (2018)
13. Jia, X., et al.: Physics-guided recurrent graph model for predicting flow and temperature in river networks. In: Proceedings of the 2021 SIAM International Conference on Data Mining (SDM), pp. 612–620. SIAM (2021)
14. Kim, D., Han, H., Wang, W., Kim, H.S.: Improvement of deep learning models for river water level prediction using complex network method. Water **14**(3), 466 (2022)
15. Zhao, Q., et al.: Joint spatial and temporal modeling for hydrological prediction. IEEE Access **8**, 78492–78503 (2020)
16. Gilmer, J., Schoenholz, S.S., Riley, P.F., Vinyals, O., Dahl, G.E.: Neural message passing for quantum chemistry. In: International Conference on Machine Learning, pp. 1263–1272. PMLR (2017)
17. Hamilton, W., Ying, Z., Leskovec, J.: Inductive representation learning on large graphs. In: Advances in Neural Information Processing Systems, vol. 30 (2017)

Author Index

A

Andersson, Axel 139

B

Bai, Lu 70
Banaeyan, Majid 160
Behanova, Andrea 139
Bigler, Vidushi 172
Brun, Luc 113

C

Carletti, Vincenzo 127
Cui, Lixin 70

D

da Cunha Cavalcanti, George Darmiton 59
de Araujo Souza, Mariana 59

E

e Cruz, Rafael Menelau Oliveira 59

F

Fadlallah, Sarah 81
Falcão, Alexandre X. 35
Fankhauser, Benjamin 172
Foggia, Pasquale 127
Fuchs, Mathias 102

G

Gaüzère, Benoît 92, 113
Gillioz, Anthony 25
Glédel, Clément 92

H

Hancock, Edwin R. 70
Honeine, Paul 92

J

Jiang, Xiaoyi 3
Julià, Carme 81

K

Kawulok, Michal 149
Kiouche, Abd Errahmane 46
Kropatsch, Walter G. 160

L

Liu, Cheng-Lin 3

M

Malmberg, Filip 35, 139
Micheli, Alessio 15

O

Ourdjini, Aymen 46

R

Riesen, Kaspar 25, 102, 172
Rossi, Luca 70

S

Sabourin, Robert 59
Seba, Hamida 46
Segura Alabart, Natália 81
Serratosa, Francesc 81
Stanovic, Stevan 113

T

Tarasiewicz, Tomasz 149
Tortorella, Domenico 15

V

Vento, Mario 127

M. Vento et al. (Eds.): GbRPR 2023, LNCS 14121, pp. 183–184, 2023.
https://doi.org/10.1007/978-3-031-42795-4

W
Wählby, Carolina   139
Wang, Yue   70

X
Xu, Zhuo   70

Y
Ye, Rongji   70

Z
Zhang, Jiaqi   3

Printed in the United States
by Baker & Taylor Publisher Services

Printed in the United States
by Baker & Taylor Publisher Services